APPLIED BIOINFORMATICS STATISTICS

APPLIED BIOINFORMATICS STATISTICS

Naresh Mahajan

RANDOM

RANDOM PUBLICATIONS

NEW DELHI - 110 002 (INDIA)

Applied Bioinformatics Statistics

ISBN 978-93-51116-92-9

Published in 2015 in India by

Reprint 2019

RANDOM PUBLICATIONS

4376-A/4B, Gali Murari Lal, Ansari Road
New Delhi-110 002
Phone: +9111-43580356, 23289044
E-mail: randomexports@gmail.com; sales@randompublications.com; info@randompublications.com

Type Setting by: Friends Media, Delhi-110089
Printed at : Mehra Printers, Delhi-110092

Preface

Bioinformatics has become an important part of many areas of biology. In experimental molecular biology, bioinformatics techniques such as image and signal processing allow extraction of useful results from large amounts of raw data. In the field of genetics and genomics, it aids in sequencing and annotating genomes and their observed mutations. It plays a role in the text mining of biological literature and the development of biological and gene ontologies to organize and query biological data. It also plays a role in the analysis of gene and protein expression and regulation. Bioinformatics tools aid in the comparison of genetic and genomic data and more generally in the understanding of evolutionary aspects of molecular biology. At a more integrative level, it helps analyze and catalogue the biological pathways and networks that are an important part of systems biology. In structural biology, it aids in the simulation and modelling of DNA, RNA, and protein structures as well as molecular interactions. In order to study how normal cellular activities are altered in different disease states, the biological data must be combined to form a comprehensive picture of these activities. Therefore, the field of bioinformatics has evolved such that the most pressing task now involves the analysis and interpretation of various types of data. This includes nucleotide and amino acid sequences, protein domains, and protein structures. The actual process of analysing and interpreting data is referred to as computational biology.

The primary goal of bioinformatics is to increase the understanding of biological processes. What sets it apart from other approaches, however, is its focus on developing and applying computationally intensive techniques to achieve this goal. Examples include: pattern recognition, data mining, machine learning algorithms, and visualization. Major research efforts in the field include sequence alignment, gene finding, genome assembly, drug design, drug discovery, protein structure alignment, protein structure prediction, prediction of gene expression and protein–protein interactions, genome-wide association studies, and the modelling of evolution. Bioinformatics now entails the creation and advancement of databases, algorithms, computational and statistical techniques, and theory to solve formal and practical problems arising from the management and analysis of biological data. Over the past few decades rapid developments in genomic and other molecular research technologies and developments in information technologies have combined to produce a tremendous amount of information related to molecular biology. Bioinformatics is the name given to these mathematical and computing approaches used to glean understanding of biological processes. Bioinformatics is a science field that is similar to but distinct from biological computation and computational biology. Biological computation uses bioengineering and biology to build biological computers, whereas bioinformatics uses computation to better understand biology. Bioinformatics and computational biology have similar aims and approaches, but they differ in scale: bioinformatics organizes and analyses basic biological data, whereas computational biology builds theoretical models of biological systems, just as mathematical biology does with mathematical models. Analysing biological data to produce meaningful information involves writing and running software programmes that use algorithms from graph theory, artificial intelligence, soft computing, data mining, image processing, and computer simulation. The algorithms in turn depend on theoretical foundations such as discrete mathematics, control theory, system theory, information theory, and statistics. The advent of modern computer technology and relatively cheap computing resources have enabled computer-intensive biostatistical methods like bootstrapping and resampling methods. Furthermore new biomedical technologies like microarrays, next generation sequencers (for genomics) and mass spectrometry (for proteomics) generate enormous amounts of data that can only be analysed with biostatistical methods.

The present book contents the foundations of key problems in computational molecular biology and bioinformatics. This book is intended to provide the comprehensive information on the field for students and researchers who are interested in the recent developments of computational statistics in computational biology.

I thank all members of my team who have helped in the preparation of the book. My special thanks go to "Random Publications" who have published the book.

— Naresh Mahajan

Contents

Chapter 1

Introduction

Bioinformatics

Bioinformatics is an interdisciplinary field that develops methods and software tools for understanding biological data. As an interdisciplinary field of science, bioinformatics combines computer science, statistics, mathematics, and engineering to study and process biological data.

Bioinformatics is both an umbrella term for the body of biological studies that use computer programming as part of their methodology, as well as a reference to specific analysis "pipelines" that are repeatedly used, particularly in the fields of genetics and genomics. Common uses of bioinformatics include the identification of candidate genes and nucleotides (SNPs). Often, such identification is made with the aim of better understanding the genetic basis of disease, unique adaptations, desirable properties (esp. in agricultural species), or differences between populations. In a less formal way, bioinformatics also tries to understand the organisational principles within nucleic acid and protein sequences.

Bioinformatics has become an important part of many areas of biology. In experimental molecular biology, bioinformatics techniques such as image and signal processing allow extraction of useful results from large amounts of raw data. In the field of genetics and genomics, it aids in sequencing and annotating genomes and their observed mutations. It plays a role in the text mining of biological literature and the development of biological and gene ontologies to organise and query biological data. It also plays a role in the analysis of gene and protein expression and regulation. Bioinformatics tools aid in the comparison of genetic and genomic data and more generally in the understanding of evolutionary aspects of molecular biology. At a more integrative level, it helps analyze and catalogue the biological pathways and networks that are an important part of systems biology. In structural biology, it aids in the simulation and modelling of DNA, RNA, and protein structures as well as molecular interactions.

History

Paulien Hogeweg and Ben Hesper coined the term *bioinformatics* in 1970 to refer to the study of information processes in biotic systems. This definition placed bioinformatics as a field parallel to biophysics (the study of physical processes in biological systems) or biochemistry (the study of chemical processes in biological systems).

Sequences

Computers became essential in molecular biology when protein sequences became available after Frederick Sanger determined the sequence of insulin in the early 1950s. Comparing multiple sequences manually turned out to be impractical. A pioneer in the field was Margaret Oakley Dayhoff, who has been hailed by David Lipman, director of the National Centre for Biotechnology Information, as the "mother and father of bioinformatics." Dayhoff compiled one of the first protein sequence databases, initially published as books and pioneered methods of sequence alignment and molecular evolution. Another early contributor to bioinformatics was Elvin A. Kabat, who pioneered biological sequence analysis in 1970 with his comprehensive volumes of antibody sequences released with Tai Te Wu between 1980 and 1991.

Genomes

As whole genome sequences became available, again with the pioneering work of Frederick Sanger, it became evident that computer-assisted analysis would be insightful. The first analysis of this type, which had important input from cryptologists at the National Security Agency, was applied to the nucleotide sequences of the bacteriophages MS2 and PhiX174. As a proof of principle, this work showed that standard methods of cryptology could reveal intrinsic features of the genetic code such as the codon length and the reading frame. This work seems to have been ahead of its time—it was rejected for publication by numerous standard journals and finally found a home in the Journal of Theoretical Biology. The term bioinformatics was re-discovered and used to refer to the creation of databases such as GenBank in 1982. With public availability of data, tools for their analysis were quickly developed and described in journals, such as

Nucleic Acids Research, which published specialised issues on bioinformatics tools as early as 1982.

Goals

To study how normal cellular activities are altered in different disease states, the biological data must be combined to form a comprehensive picture of these activities. Therefore, the field of bioinformatics has evolved such that the most pressing task now involves the analysis and interpretation of various types of data. This includes nucleotide and amino acid sequences, protein domains, and protein structures. The actual process of analyzing and interpreting data is referred to as computational biology. Important sub-disciplines within bioinformatics and computational biology include:

- Development and implementation of computer programs that enable efficient access to, use and management of, various types of information
- Development of new algorithms (mathematical formulas) and statistical measures that assess relationships among members of large data sets. For example, there are methods to locate a gene within a sequence, to predict protein structure and/or function, and to cluster protein sequences into families of related sequences.

The primary goal of bioinformatics is to increase the understanding of biological processes. What sets it apart from other approaches, however, is its focus on developing and applying computationally intensive techniques to achieve this goal. Examples include: pattern recognition, data mining, machine learning algorithms, and visualization. Major research efforts in the field include sequence alignment, gene finding, genome assembly, drug design, drug discovery, protein structure alignment, protein structure prediction, prediction of gene expression and protein–protein interactions, genome-wide association studies, and the modelling of evolution.

Bioinformatics now entails the creation and advancement of databases, algorithms, computational and statistical techniques, and theory to solve formal and practical problems arising from the management and analysis of biological data.

Over the past few decades rapid developments in genomic and other molecular research technologies and developments in information technologies have combined to produce a tremendous amount of information related to molecular biology. Bioinformatics is the name given to these mathematical and computing approaches used to glean understanding of biological processes.

Approaches

Common activities in bioinformatics include mapping and analyzing DNA and protein sequences, aligning DNA and protein sequences to compare them, and creating and viewing 3-D models of protein structures.

There are two fundamental ways of modelling a Biological system (e.g., living cell) both coming under Bioinformatic approaches.

- Static
  - o Sequences – Proteins, Nucleic acids and Peptides
  - o Interaction data among the above entities including microarray data and Networks of proteins, metabolites
- Dynamic
  - o Structures – Proteins, Nucleic acids, Ligands (including metabolites and drugs) and Peptides (structures studied with bioinformatics tools are not considered static anymore and their dynamics is often the core of the structural studies)
  - o Systems Biology comes under this category including reaction fluxes and variable concentrations of metabolites
  - o Multi-Agent Based modelling approaches capturing cellular events such as signalling, transcription and reaction dynamics

A broad sub-category under bioinformatics is structural bioinformatics.

Relation to Other Fields

Bioinformatics is a science field that is similar to but distinct from biological computation and computational biology. Biological computation uses bioengineering and biology to build biological computers, whereas bioinformatics uses computation to better understand biology. Bioinformatics and computational biology have similar aims and approaches, but they differ in scale: bioinformatics organises and analyzes basic biological data, whereas computational biology builds theoretical models of biological systems, just as mathematical biology does with mathematical models.

Analyzing biological data to produce meaningful information involves writing and running software programs that use algorithms from graph theory, artificial intelligence, soft computing, data mining, image processing, and computer simulation. The algorithms in turn depend on theoretical foundations such as discrete mathematics, control theory, system theory, information theory, and statistics.

Sequence Analysis

Figure: *The sequences of different genes or proteins may be aligned side-by-side to measure their similarity. This alignment compares protein sequences containing WPP domains.*

Since the Phage ϑ-X174 was sequenced in 1977, the DNA sequences of thousands of organisms have been decoded and stored in databases. This sequence information is analyzed to determine genes that encode proteins, RNA genes, regulatory sequences, structural motifs, and repetitive sequences. A comparison of genes within a species or between different species can show similarities between protein functions, or relations between species (the use of molecular systematics to construct phylogenetic trees). With the growing amount of data, it long ago became impractical to analyze DNA sequences manually. Today, computer programs such as BLAST are used daily to search sequences from more than 260 000 organisms, containing over 190 billion nucleotides. These programs can compensate for mutations (exchanged, deleted or inserted bases) in the DNA sequence, to identify sequences that are related, but not identical. A variant of this sequence alignment is used in the sequencing process itself.

The so-called shotgun sequencing technique (which was used, for example, by The Institute for Genomic Research to sequence the first bacterial genome, *Haemophilus influenzae*) does not produce entire chromosomes. Instead it generates the sequences of many thousands of small DNA fragments (ranging from 35 to 900 nucleotides long, depending on the sequencing technology). The ends of these fragments overlap and, when aligned properly by a genome assembly program, can be used to reconstruct the complete genome. Shotgun sequencing yields sequence data quickly, but the task of assembling the fragments can be quite complicated for larger genomes. For a genome as large as the human genome, it may take many days of CPU time on large-memory, multiprocessor computers to assemble the fragments, and the resulting assembly usually contains numerous gaps that must be filled in later. Shotgun sequencing is the method of choice for virtually all genomes sequenced today, and genome assembly algorithms are a critical area of bioinformatics research.

Another aspect of bioinformatics in sequence analysis is annotation. This involves computational gene finding to search for protein-coding genes, RNA genes, and other functional sequences within a genome. Not all of the nucleotides within a genome are part of genes. Within the genomes of higher organisms, large parts of the DNA do not serve any obvious purpose. This so-called junk DNA may, however, contain unrecognised functional elements. Bioinformatics helps to bridge the gap between genome and proteome projects — for example, in the use of DNA sequences for protein identification.

Genome Annotation

In the context of genomics, annotation is the process of marking the genes and other biological features in a DNA sequence. This process needs to be automated because most genomes are too large to annotate by hand, not to mention the desire to annotate as many genomes as possible, as the rate of sequencing has ceased to pose a bottleneck. Annotation is made possible by the fact that genes have recognisable start and stop regions, although the exact sequence found in these regions can vary between genes.

The first genome annotation software system was designed in 1995 by Owen White, who was part of the team at The Institute for Genomic Research that sequenced and analyzed the first genome of a free-living organism to be decoded, the bacterium *Haemophilus influenzae*. White built a software system to find the genes (fragments of genomic sequence that encode proteins), the transfer RNAs, and to make initial assignments of function to those genes. Most current genome annotation systems work similarly, but the programs available for analysis of genomic DNA, such as the GeneMark program trained and used to find protein-coding genes in *Haemophilus influenzae*, are constantly changing and improving.

Computational Evolutionary Biology

Evolutionary biology is the study of the origin and descent of species, as well as their change over time. Informatics has assisted evolutionary biologists by enabling researchers to:

- trace the evolution of a large number of organisms by measuring changes in their DNA, rather than through physical taxonomy or physiological observations alone,
- more recently, compare entire genomes, which permits the study of more complex evolutionary events, such as gene duplication, horizontal gene transfer, and the prediction of factors important in bacterial speciation,

- build complex computational models of populations to predict the outcome of the system over time
- track and share information on an increasingly large number of species and organisms

Future work endeavours to reconstruct the now more complex tree of life.

The area of research within computer science that uses genetic algorithms is sometimes confused with computational evolutionary biology, but the two areas are not necessarily related.

Comparative Genomics

The core of comparative genome analysis is the establishment of the correspondence between genes (orthology analysis) or other genomic features in different organisms. It is these intergenomic maps that make it possible to trace the evolutionary processes responsible for the divergence of two genomes. A multitude of evolutionary events acting at various organisational levels shape genome evolution. At the lowest level, point mutations affect individual nucleotides. At a higher level, large chromosomal segments undergo duplication, lateral transfer, inversion, transposition, deletion and insertion. Ultimately, whole genomes are involved in processes of hybridization, polyploidization and endosymbiosis, often leading to rapid speciation. The complexity of genome evolution poses many exciting challenges to developers of mathematical models and algorithms, who have recourse to a spectra of algorithmic, statistical and mathematical techniques, ranging from exact, heuristics, fixed parameter and approximation algorithms for problems based on parsimony models to Markov Chain Monte Carlo algorithms for Bayesian analysis of problems based on probabilistic models.

Many of these studies are based on the homology detection and protein families computation.

Genetics of Disease

With the advent of next-generation sequencing we are obtaining enough sequence data to map the genes of complex diseases such as infertility, breast cancer or Alzheimer's Disease. Genome-wide association studies are essential to pinpoint the mutations for such complex diseases.

Analysis of Mutations in Cancer

In cancer, the genomes of affected cells are rearranged in complex or even unpredictable ways. Massive sequencing efforts are used to identify previously unknown point mutations in a variety of genes in cancer. Bioinformaticians continue to produce specialised automated systems to manage the sheer volume of sequence data produced, and they create new algorithms and software to compare the sequencing results to the growing collection of human genome sequences and germline polymorphisms. New physical detection technologies are employed, such as oligonucleotide microarrays to identify chromosomal gains and losses (called comparative genomic hybridization), and single-nucleotide polymorphism arrays to detect known *point mutations*. These detection methods simultaneously measure several hundred thousand sites throughout the genome, and when used in high-throughput to measure thousands of samples, generate terabytes of data per experiment. Again the massive amounts and new types of data generate new opportunities for bioinformaticians. The data is often found to contain considerable variability, or noise, and thus Hidden Markov model and change-point analysis methods are being developed to infer real copy number changes.

Another type of data that requires novel informatics development is the analysis of lesions found to be recurrent among many tumors.

Gene and Protein Expression

Analysis of Gene Expression

The expression of many genes can be determined by measuring mRNA levels with multiple techniques including microarrays, expressed cDNA sequence tag (EST) sequencing, serial analysis of gene expression (SAGE) tag sequencing, massively parallel signature sequencing (MPSS), RNA-Seq, also known as "Whole Transcriptome Shotgun Sequencing" (WTSS), or various applications of multiplexed in-situ hybridization. All of these techniques are extremely noise-prone and/or subject to bias in the biological measurement, and a major research area in computational biology involves developing statistical tools to separate signal from noise in high-throughput gene expression studies. Such studies are often used to determine the genes implicated in a disorder: one might compare microarray data from cancerous epithelial cells to data from non-cancerous cells to determine the transcripts that are up-regulated and down-regulated in a particular population of cancer cells.

Analysis of Protein Expression

Protein microarrays and high throughput (HT) mass spectrometry (MS) can provide a snapshot of the proteins present in a biological sample. Bioinformatics is very much involved in making sense of protein microarray and HT MS data; the former approach faces similar problems as with microarrays targeted at mRNA, the latter involves the problem of matching large amounts of mass data against predicted masses from protein sequence databases, and the complicated statistical analysis of samples where multiple, but incomplete peptides from each protein are detected.

Analysis of Regulation

Regulation is the complex orchestration of events starting with an extracellular signal such as a hormone and leading to an increase or decrease in the activity of one or more proteins. Bioinformatics techniques have been applied to explore various steps in this process. For example, promoter analysis involves the identification and study of sequence motifs in the DNA surrounding the coding region of a gene. These motifs influence the extent to which that region is transcribed into mRNA. Expression data can be used to infer gene regulation: one might compare microarray data from a wide variety of states of an organism to form hypotheses about the genes involved in each state. In a single-cell organism, one might compare stages of the cell cycle, along with various stress conditions (heat shock, starvation, etc.). One can then apply clustering algorithms to that expression data to determine which genes are co-expressed. For example, the upstream regions (promoters) of co-expressed genes can be searched for over-represented regulatory elements. Examples of clustering algorithms applied in gene clustering are k-means clustering, self-organising maps (SOMs), hierarchical clustering, and consensus clustering methods such as the Bi-CoPaM. The later, namely Bi-CoPaM, has been actually proposed to address various issues specific to gene discovery problems such as consistent co-expression of genes over multiple microarray datasets.

Structural Bioinformatics

Protein structure prediction is another important application of bioinformatics. The amino acid sequence of a protein, the so-called primary structure, can be easily determined from the sequence on the gene that codes for it. In the vast majority of cases, this primary structure uniquely determines a structure in its native environment. (Of course, there are exceptions, such as the bovine spongiform encephalopathy – a.k.a. Mad Cow Disease – prion.) Knowledge of this structure is vital in understanding the function of the protein. Structural information is usually classified as one of *secondary*, *tertiary* and *quaternary* structure. A viable general solution to such predictions remains an open problem. Most efforts have so far been directed towards heuristics that work most of the time.

One of the key ideas in bioinformatics is the notion of homology. In the genomic branch of bioinformatics, homology is used to predict the function of a gene: if the sequence of gene *A*, whose function is known, is homologous to the sequence of gene *B*, whose function is unknown, one could infer that B may share A's function. In the structural branch of bioinformatics, homology is used to determine which parts of a protein are important in structure formation and interaction with other proteins. In a technique called homology modelling, this information is used to predict the structure of a protein once the structure of a homologous protein is known. This currently remains the only way to predict protein structures reliably.

One example of this is the similar protein homology between hemoglobin in humans and the hemoglobin in legumes (leghemoglobin). Both serve the same purpose of transporting oxygen in the organism. Though both of these proteins have completely different amino acid sequences, their protein structures are virtually identical, which reflects their near identical purposes.

Other techniques for predicting protein structure include protein threading and *de novo* (from scratch) physics-based modelling.

Network and Systems Biology

Network analysis seeks to understand the relationships within biological networks such as metabolic or protein-protein interaction networks. Although biological networks can be constructed from a single type of molecule or entity (such as genes), network biology often attempts to integrate many different data types, such as proteins, small molecules, gene expression data, and others, which are all connected physically, functionally, or both.

Systems biology involves the use of computer simulations of cellular subsystems (such as the networks of metabolites and enzymes that comprise metabolism, signal transduction pathways and gene regulatory networks) to both analyze and visualize the complex connections of these cellular processes. Artificial life or virtual evolution attempts to understand evolutionary processes via the computer simulation of simple (artificial) life forms.

Molecular Interaction Networks

Tens of thousands of three-dimensional protein structures have been determined by X-ray crystallography and protein nuclear magnetic resonance spectroscopy (protein NMR) and a central question in structural bioinformatics is whether it is practical to predict possible protein–protein interactions only based on these 3D shapes, without performing protein–protein interaction experiments. A variety of methods have been developed to tackle the protein–protein docking problem, though it seems that there is still much work to be done in this field.

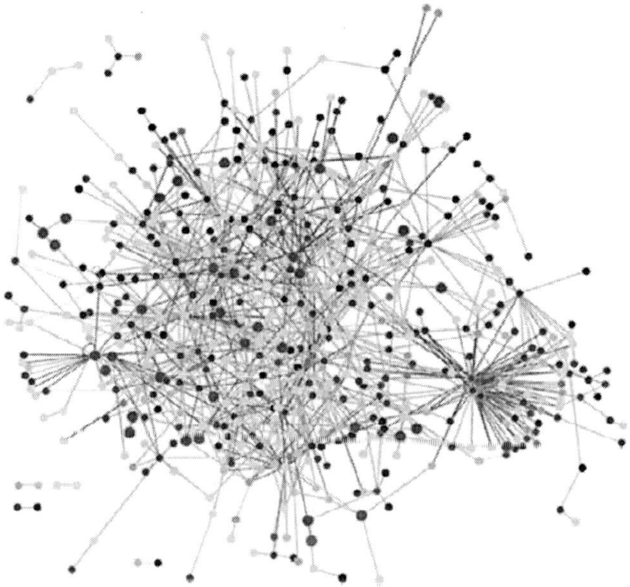

Figure: *Interactions between proteins are frequently visualized and analyzed using networks. This network is made up of protein-protein interactions from* Treponema pallidum, *the causative agent of syphilis and other diseases.*

Other interactions encountered in the field include Protein–ligand (including drug) and protein–peptide. Molecular dynamic simulation of movement of atoms about rotatable bonds is the fundamental principle behind computational algorithms, termed docking algorithms, for studying molecular interactions.

Others

Literature Analysis

The growth in the number of published literature makes it virtually impossible to read every paper, resulting in disjointed sub-fields of research. Literature analysis aims to employ computational and statistical linguistics to mine this growing library of text resources. For example:

- Abbreviation recognition – identify the long-form and abbreviation of biological terms
- Named entity recognition – recognising biological terms such as gene names
- Protein-protein interaction – identify which proteins interact with which proteins from text

The area of research draws from statistics and computational linguistics.

High-Throughput Image Analysis

Computational technologies are used to accelerate or fully automate the processing, quantification and analysis of large amounts of high-information-content biomedical imagery. Modern image analysis systems augment an observer's ability to make measurements from a large or complex set of images, by improving accuracy, objectivity, or speed. A fully developed analysis system may completely replace the observer.

Although these systems are not unique to biomedical imagery, biomedical imaging is becoming more important for both diagnostics and research. Some examples are:

- high-throughput and high-fidelity quantification and sub-cellular localization (high-content screening, cytohistopathology, Bioimage informatics)
- morphometrics
- clinical image analysis and visualization
- determining the real-time air-flow patterns in breathing lungs of living animals
- quantifying occlusion size in real-time imagery from the development of and recovery during arterial injury
- making behavioural observations from extended video recordings of laboratory animals
- infrared measurements for metabolic activity determination
- inferring clone overlaps in DNA mapping, e.g. the Sulston score

High-Throughput Single Cell Data Analysis

Computational techniques are used to analyse high-throughput, low-measurement single cell data, such as that obtained from flow cytometry. These methods typically involve finding populations of cells that are relevant to a particular disease state or experimental condition.

Biodiversity Informatics

Biodiversity informatics deals with the collection and analysis of biodiversity data, such as taxonomic databases, or microbiome data. Examples of such analyses include phylogenetics, niche modelling, species richness mapping, or species identification tools.

Databases

Databases are essential for bioinformatics research and applications. There is a huge number of available databases covering almost everything from DNA and protein sequences, molecular structures, to phenotypes and biodiversity. Databases generally fall into one of three types. Some contain data resulting directly from empirical methods such as gene knockouts. Others consist of predicted data, and most contain data from both sources. There are meta-databases that incorporate data compiled from multiple other databases. Some others are specialised, such as those specific to an organism. These databases vary in their format, way of accession and whether they are public or not. Some of the most commonly used databases are listed below. For a more comprehensive list, please check the link at the beginning of the subsection.

- Used in Motif Finding: GenomeNet MOTIF Search
- Used in Gene Ontology: DAVID, FuncAssociate, GATHER
- Used in Gene Finding: Hidden Markov Model
- Used in finding Protein Structures/Family: PFAM
- Used for Next Generation Sequencing: (Not database but data format), FASTQ Format
- Used in Gene Expression Analysis: GEO
- Used in Network Analysis: Interaction Analysis Databases(BioGRID, MINT, HPRD, Curated Human Signaling Network), Functional Networks (STRING, KEGG)

Please keep in mind that this is a quick sampling and generally most computation data is supported by wet lab data as well.

Software and Tools

Software tools for bioinformatics range from simple command-line tools, to more complex graphical programs and standalone web-services available from various bioinformatics companies or public institutions.

Open-Source Bioinformatics Software

Many free and open-source software tools have existed and continued to grow since the 1980s. The combination of a continued need for new algorithms for the analysis of emerging types of biological readouts, the potential for innovative *in silico* experiments, and freely available open code bases have helped to create opportunities for all research groups to contribute to both bioinformatics and the range of open-source software available, regardless of their funding arrangements. The open source tools often act as incubators of ideas, or community-supported plug-ins in commercial applications. They may also provide *de facto* standards and shared object models for assisting with the challenge of bioinformation integration.

The range of open-source software packages includes titles such as Bioconductor, BioPerl, Biopython, BioJava, BioJS, BioRuby, Bioclipse, EMBOSS, .NET Bio, Taverna workbench, and UGENE. To maintain this tradition and create further opportunities, the non-profit Open Bioinformatics Foundation have supported the annual Bioinformatics Open Source Conference (BOSC) since 2000.

Web Services in Bioinformatics

SOAP- and REST-based interfaces have been developed for a wide variety of bioinformatics applications allowing an application running on one computer in one part of the world to use algorithms, data and computing resources on servers in other parts of the world. The main advantages derive from the fact that end users do not have to deal with software and database maintenance overheads.

Basic bioinformatics services are classified by the EBI into three categories: SSS (Sequence Search Services), MSA (Multiple Sequence Alignment), and BSA (Biological Sequence Analysis). The availability of these service-oriented bioinformatics resources demonstrate the applicability of web-based bioinformatics solutions, and range from a collection of standalone tools with a common data format under a single, standalone or web-based interface, to integrative, distributed and extensible bioinformatics workflow management systems.

Bioinformatics Workflow Management Systems

A Bioinformatics workflow management system is a specialised form of a workflow management system designed specifically to compose and execute a series of computational or data manipulation steps, or a workflow, in a Bioinformatics application. Such systems are designed to

- provide an easy-to-use environment for individual application scientists themselves to create their own workflows
- provide interactive tools for the scientists enabling them to execute their workflows and view their results in real-time

- simplify the process of sharing and reusing workflows between the scientists.
- enable scientists to track the provenance of the workflow execution results and the workflow creation steps.

Some of the platforms giving this service: Galaxy, Kepler, Taverna, UGENE, Anduril.

Sequence Database

In the field of bioinformatics, a sequence database is a type of biological database that is composed of a large collection of computerized ("digital") nucleic acid sequences, protein sequences, or other polymer sequences stored on a computer.

The UniProt database is an example of a protein sequence database. As of 2013 it contained over 40 million sequences and is growing at an exponential rate. Historically, sequences were published in paper form, but as the number of sequences grew this storage method became unsustainable.

Search Issues

Sequence databases can be searched using a variety of methods. The most common usage is probably searching for sequences similar to a certain target protein or gene whose sequence is already known to the user. The BLAST program is a popular method of this type.

Current Issues

Records in sequence databases are deposited from a wide range of sources, from individual researchers to large genome sequencing centres. As a result, the sequences themselves, and especially the biological annotations attached to these sequences, may vary in quality. There is much redundancy, as multiple labs may submit numerous sequences that are identical, or nearly identical, to others in the databases.

Many annotations of the sequences are based not on laboratory experiments, but on the results of sequence similarity searches for previously-annotated sequences. Once a sequence has been annotated based on similarity to others, and itself deposited in the database, it can also become the basis for future annotations.

This can lead to a *transitive annotation problem* because there may be several such annotation transfers by sequence similarity between a particular database record and actual wet lab experimental information. Therefore, care must be taken when interpreting the annotation data from sequence databases.

Protein Structure Database

In biology, a protein structure database is a database that is modelled around the various experimentally determined protein structures. The aim of most protein structure databases is to organise and annotate the protein structures, providing the biological community access to the experimental data in a useful way. Data included in protein structure databases often includes three-dimensional coordinates as well as experimental information, such as unit cell dimensions and angles for x-ray crystallography determined structures.

Though most instances, in this case either proteins or a specific structure determinations of a protein, also contain sequence information and some databases even provide means for performing sequence based queries, the primary attribute of a structure database is structural information, whereas sequence databases focus on sequence information, and contain no structural information for the majority of entries. Protein structure databases are critical for many efforts in computational biology such as structure based drug design, both in developing the computational methods used and in providing a large experimental dataset used by some methods to provide insights about the function of a protein.

The Protein Data Bank

The Protein Data Bank (PDB) was established in 1971 as the central archive of all experimentally determined protein structure data. Today the PDB is maintained by an international consortia collectively known as the Worldwide Protein Data Bank (wwPDB). The mission of the wwPDB is to maintain a single archive of macromolecular structural data that is freely and publicly available to the global community.

List of Other Protein Structure Databases

Because the PDB releases data into the public domain, the data has been used in various other protein structure databases.

Examples of protein structure databases include (in alphabetical order);

Database of Macromolecular: Movements describes the motions that occur in proteins and other macromolecules, particularly using movies

Dynameomics: a data warehouse of molecular dynamics simulations and analyses of proteins representing all known protein fold families

Jena: Lib the Jena Library of Biological Macromolecules is aimed at a better dissemination of information on three-dimensional biopolymer structures with an emphasis on visualization and analysis.

Mod: Base a database of three-dimensional protein models calculated by comparative modelling

OCA: a browser-database for protein structure/function - The OCA integrates information from KEGG, OMIM, PDBselect, Pfam, PubMed, SCOP, SwissProt, and others.

OPM: provides spatial positions of protein three-dimensional structures with respect to the lipid bilayer.

PDBe: the European resource for the collection, organisation and dissemination of data on biological macromolecular structures, and a member of the Worldwide Protein Data Bank.

PDB: Lite derived from OCA, PDB Lite was provided to make it as easy as possible to find and view a macromolecule within the PDB

PDB: sum provides an overview macromolecular structures in the PDB, giving schematic diagrams of the molecules in each structure and of the interactions between them

PDBTM: the Protein Data Bank of Transmembrane Proteins — a selection of the PDB.

PDB: Wiki a community annotated knowledge base of biological molecular structures.

Prot: CID The Protein Common Interface Database (ProtCID) is a database of similar protein-protein interfaces in crystal structures of homologous proteins.

Protein the NIH protein database, a collection of sequences from several sources, including translations from annotated coding regions in GenBank, RefSeq and Third Party Annotation, as well as records from SwissProt, PIR, PRF, and PDB

Proteopedia the collaborative, 3D encyclopedia of proteins and other molecules. A wiki that contains a page for every entry in the PDB (>100,000 pages), with a Jmol view that highlights functional sites and ligands. Offers an easy-to-use scene-authoring tool so you don't have to learn Jmol script language to create customized molecular scenes. Custom scenes are easily attached to "green links" in descriptive text that display those scenes in Jmol.

Protein: Lounge a protein databases that includes visuals of protein structure. Also, includes protein pathways and gene sequences including other tools.

SCOP: the Structural Classification of Proteins [2] a detailed and comprehensive description of the structural and evolutionary relationships between all proteins whose structure is known.

SWISS-MODEL Repository: a database of annotated protein models calculated by homology modelling

TOPSAN: the Open Protein Structure Annotation Network — a wiki designed to collect, share and distribute information about protein three-dimensional structures.

Gene Prediction

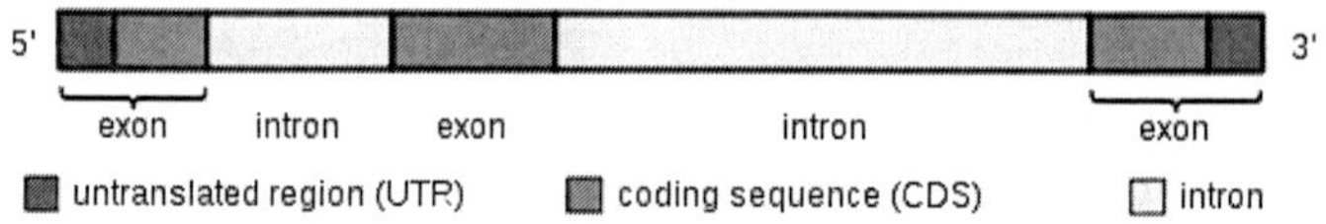

Figure: *Structure of a gene*

In computational biology gene prediction or gene finding refers to the process of identifying the regions of genomic DNA that encode genes. This includes protein-coding genes as well as RNA genes, but may also include prediction of other functional elements such as regulatory regions. Gene finding is one of the first and most important steps in understanding the genome of a species once it has been sequenced.

In its earliest days, "gene finding" was based on painstaking experimentation on living cells and organisms. Statistical analysis of the rates of homologous recombination of several different genes could determine their order on a certain chromosome, and information from many such experiments could be combined to create a genetic map specifying the rough location of known genes relative to each other. Today, with comprehensive genome sequence and powerful computational resources at the disposal of the research community, gene finding has been redefined as a largely computational problem.

Determining that a sequence is functional should be distinguished from determining the function of the gene or its product. Predicting the function of a gene and confirming that the gene prediction is accurate still demands *in vivo* experimentation through gene knockout and other assays, although frontiers of bioinformatics research are making it increasingly possible to predict the function of a gene based on its sequence alone.

Gene prediction is one of the key steps in Genome annotation, following Sequence assembly, the filtering of non-coding regions and repeat masking.

Many aspects of structural gene prediction are based on current understanding of underlying biochemical processes in the cell such as gene transcription, translation, protein–protein interactions and regulation processes, which are subject of active research in the various Omics fields such as Transcriptomics, Proteomics, Metabolomics, and more generally structural and functional genomics.

Empirical Methods

In empirical (similarity, homology or evidence-based) gene finding systems, the target genome is searched for sequences that are similar to extrinsic evidence in the form of the known expressed sequence tags, messenger RNA (mRNA), protein products, and homologous or orthologous sequences. Given an mRNA sequence, it is trivial to derive a unique genomic DNA sequence from which it had to have been transcribed. Given a protein sequence, a family of possible coding DNA sequences can be derived by reverse translation of the genetic code. Once candidate DNA sequences have been determined, it is a relatively straightforward algorithmic problem to efficiently search a target genome for matches, complete or partial, and exact or inexact. Given a sequence, local alignment algorithms such as BLAST, FASTA and Smith-Waterman look for regions of similarity between the target sequence and possible candidate matches. Matches can be complete or partial, and exact or inexact. The success of this approach will be limited by the contents and accuracy of the sequence database.

A high degree of similarity to a known messenger RNA or protein product is strong evidence that a region of a target genome is a protein-coding gene. However, to apply this approach systemically requires extensive sequencing of mRNA and protein products. Not only is this expensive, but in complex organisms, only a subset of all genes in the organism's genome are expressed at any given time, meaning that extrinsic evidence for many genes is not readily accessible in any single cell culture. Thus, in order to collect extrinsic evidence for most or all of the genes in a complex organism, many hundreds or thousands of different cell types must be studied, which itself presents further difficulties. For example, some human genes may be expressed only during development as an embryo or fetus, which might be difficult to study for ethical reasons.

Despite these difficulties, extensive transcript and protein sequence databases have been generated for human as well as other important model organisms in biology, such as mice and yeast. For example, the RefSeq database contains transcript and protein sequence from many different species, and the Ensembl system comprehensively maps this evidence to human and several other genomes. It is, however, likely that these databases are both incomplete and contain small but significant amounts of erroneous data.

New high-throughput Transcriptome sequencing technologies such as RNA-Seq and ChIP-sequencing open opportunities for incorporating additional extrinsic evidence into gene prediction and validation, and allow structurally rich and more accurate alternative to previous methods of measuring Gene expression such as Expressed sequence tag or DNA microarray.

Major challenges involved in gene prediction involve dealing with sequencing errors in raw DNA data, dependence on the quality of the Sequence assembly, handling short reads, Frameshift mutations, Overlapping genes and incomplete genes.

In prokaryotes it's essential to consider Horizontal gene transfer when searching for gene sequence homology. An additional important factor which is somewhat underutilised in current gene detection tools is existence of gene clusters - Operons in both prokaryotes and eukaryotes. Most of the popular gene detectors treat each gene in isolation, independent of others which is not biologically accurate.

Ab Initio Methods

Ab Initio gene prediction is an intrinsic method based on gene content and signal detection. Because of the inherent expense and difficulty in obtaining extrinsic evidence for many genes, it is also necessary to resort to *Ab initio* gene finding, in which genomic DNA sequence alone is systematically searched for certain tell-tale signs of protein-coding genes. These signs can be broadly categorized as either *signals*, specific sequences that indicate the presence of a gene nearby, or *content*, statistical properties of protein-coding sequence itself. *Ab initio* gene finding might be more accurately characterized as gene *prediction*, since extrinsic evidence is generally required to conclusively establish that a putative gene is functional.

In the genomes of prokaryotes, genes have specific and relatively well-understood promoter sequences (signals), such as the Pribnow box and transcription factor binding sites, which are easy to systematically identify. Also, the sequence coding for a protein occurs as one contiguous open reading frame (ORF), which is typically many hundred or thousands of base pairs long. The statistics of stop codons are such that even finding an open reading frame of this length is a fairly informative sign. (Since 3 of the 64 possible codons in the genetic code are stop codons, one would expect a stop codon approximately every

20–25 codons, or 60–75 base pairs, in a random sequence.) Furthermore, protein-coding DNA has certain periodicities and other statistical properties that are easy to detect in sequence of this length. These characteristics make prokaryotic gene finding relatively straightforward, and well-designed systems are able to achieve high levels of accuracy.

Ab initio gene finding in eukaryotes, especially complex organisms like humans, is considerably more challenging for several reasons. First, the promoter and other regulatory signals in these genomes are more complex and less well-understood than in prokaryotes, making them more difficult to reliably recognise. Two classic examples of signals identified by eukaryotic gene finders are CpG islands and binding sites for a poly(A) tail.

Second, splicing mechanisms employed by eukaryotic cells mean that a particular protein-coding sequence in the genome is divided into several parts (exons), separated by non-coding sequences (introns). (Splice sites are themselves another signal that eukaryotic gene finders are often designed to identify.) A typical protein-coding gene in humans might be divided into a dozen exons, each less than two hundred base pairs in length, and some as short as twenty to thirty. It is therefore much more difficult to detect periodicities and other known content properties of protein-coding DNA in eukaryotes.

Advanced gene finders for both prokaryotic and eukaryotic genomes typically use complex probabilistic models, such as hidden Markov models (HMMs), in order to combine information from a variety of different signal and content measurements. The GLIMMER system is a widely used and highly accurate gene finder for prokaryotes. GeneMark is another popular approach. Eukaryotic *ab initio* gene finders, by comparison, have achieved only limited success; notable examples are the GENSCAN and geneid programs. The SNAP gene finder is HMM-based like Genscan and attempts to be more adaptable to different organisms, addressing problems related to using a gene finder on a genome sequence that it was not trained against. A few recent approaches like mSplicer, CONTRAST, or mGene also use machine learning techniques like support vector machines for successful gene prediction. They build a discriminative model using hidden Markov support vector machines or conditional random fields to learn an accurate gene prediction scoring function.

Ab Initio methods have been benchmarked, with some approaching 100% sensitivity, however as the sensitivity increases, accuracy suffers as a result of increased false positives.

Other Signals

Among the derived signals used for prediction are statistics resulting from the sub-sequence statistics like k-mer statistics, Isochore (genetics) or Compositional domain GC composition/uniformity/entropy, sequence and frame length, Intron/Exon/Donor/Acceptor/Promoter and Ribosomal binding site vocabulary, Fractal dimension, Fourier transform of a pseudo-number-coded DNA, Z-curve parameters and certain run features.

It has been suggested that signals other than those directly detectable in sequences may improve gene prediction. For example, the role of secondary structure in the identification of regulatory motifs has been reported. In addition, it has been suggested that RNA secondary structure prediction helps splice site prediction.

Neural Networks

Neural networks are computational models which excel at machine learning and pattern recognition. Neural networks must be trained with example data before being able to generalise for experimental data, and tested against benchmark data. Neural networks are able to come up with approximate solutions to problems which are hard to solve algorithmically, provided there is sufficient training data. When applied to gene prediction, neural networks can be used alongside other *ab initio* methods to predict or identify biological features such as splice sites. One approach involves using a sliding window, which traverses the sequence data in an overlapping manner. The output at each position is a score based on whether the network thinks the window contains a donor splice site or an acceptor splice site. Larger windows offer more accuracy but also require more computational power. A neural network is an example of a signal sensor as its goal is to identify a functional site in the genome.

Combined Approaches

Programs such as Maker combine extrinsic and *ab initio* approaches by mapping protein and EST data to the genome to validate *ab initio* predictions. Augustus, which may be used as part of the Maker pipeline, can also incorporate hints in the form of EST alignments or protein profiles to increase the accuracy of the gene prediction.

Comparative Genomics Approaches

As the entire genomes of many different species are sequenced, a promising direction in current research on gene finding is a comparative genomics approach.

This is based on the principle that the forces of natural selection cause genes and other functional elements to undergo mutation at a slower rate than the rest of the genome, since mutations in functional elements are more likely to negatively impact the organism than mutations elsewhere. Genes can thus be detected by comparing the genomes of related species to detect this evolutionary pressure for conservation. This approach was first applied to the mouse and human genomes, using programs such as SLAM, SGP and TWINSCAN/N-SCAN and CONTRAST.

Multiple Informants

TWINSCAN examined only human-mouse synteny to look for orthologous genes. Programs such as N-SCAN and CONTRAST allowed the incorporation of alignments from multiple organisms, or in the case of N-SCAN, a single alternate organism from the target. The use of multiple informants can lead to significant improvements in accuracy.

CONTRAST is composed of two elements. The first is a smaller classifier, identifying donor splice sites and acceptor splice sites as well as start and stop codons. The second element involves constructing a full model using machine learning. Breaking the problem into two means that smaller targeted data sets can be used to train the classifiers, and that classifier can operate independently and be trained with smaller windows.

The full model can use the independent classifier, and not have to waste computational time or model complexity re-classifying intron-exon boundaries. The paper in which CONTRAST is introduced proposes that their method (and those of TWINSCAN, etc.) be classified as *de novo* gene assembly, using alternate genomes, and identifying it as distinct from *ab initio*, which uses a target 'informant' genomes.

Comparative gene finding can also be used to project high quality annotations from one genome to another. Notable examples include Projector, GeneWise and GeneMapper. Such techniques now play a central role in the annotation of all genomes.

Pseudogene Prediction

Pseudogenes are close relatives of genes, sharing very high sequence homology, but being unable to code for the same protein product. Whilst once relegated as byproducts of gene sequencing, increasingly, as regulatory roles are being uncovered, they are becoming predictive targets in their own right. Pseudogene prediction utilises existing sequence similarity and ab initio methods, whilst adding additional filtering and methods of identifying pseudogene characteristics.

Sequence similarity methods can be customised for pseudogene prediction using additional filtering to find candidate pseudogenes. This could use disablement detection, which looks for nonsense or frameshift mutations which would truncate or collapse an otherwise functional coding sequence. Additionally, translating DNA into proteins sequences can be more effective than just straight DNA homology.

Content sensors can be filtered according to the differences in statistical properties between pseudogenes and genes, such as a reduced count of CpG islands in pseudogenes, or the differences in G-C content between pseudogenes and their neighbours. Signal sensors also can be honed to pseudogenes, looking for the absence of introns or polyadenine tails.

Metagenomic Gene Prediction

Metagenomics is the study of genetic material recovered from the environment, resulting in sequence information from a pool of organisms. Predicting genes is useful for comparative metagenomics.

Metagenomics tools also fall into the basic categories of using either sequence similarity approaches (MEGAN4) and ab initio techniques (GLIMMER-MG).

Glimmer-MG is an extension to GLIMMER that relies mostly on an ab initio approach for gene finding and by using training sets from related organisms. The prediction strategy is augmented by classification and clustering gene data sets prior to applying ab initio gene prediction methods. The data is clustered by species. This classification method leverages techniques from metagenomic phylogenetic classification. An example of software for this purpose is, Phymm, which uses interpolated markov models, and PymmBL which integrates BLAST into the classification routines.

MEGAN4 uses a sequence similarity approach, using local alignment against databases of known sequences, but also attempts to classify using additional information on functional roles, biological pathways and enzymes. As in single organism gene prediction, sequence similarity approaches are limited by the size of the database.

Frag GeneScan and MetaGeneAnnotator are popular gene prediction programs based on Hidden Markov model. These predictors account for sequencing errors, partial genes and work for short reads.

Sequence Motif

In genetics, a sequence motif is a nucleotide or amino-acid sequence pattern that is widespread and has, or is conjectured to have, a biological significance. For proteins, a sequence motif is distinguished from a structural motif, a motif formed by the three-dimensional arrangement of amino acids, which may not be adjacent.

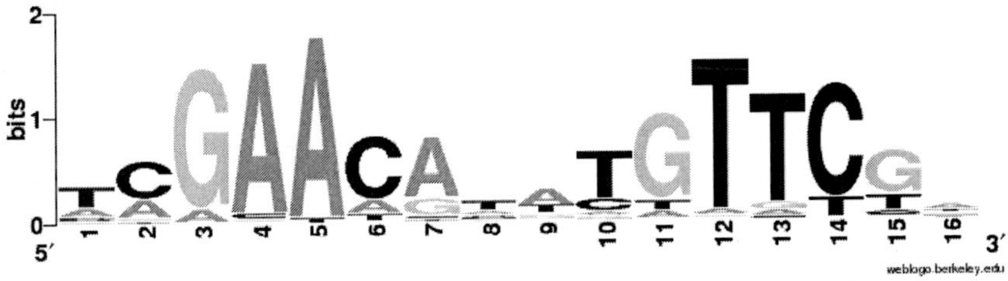

Figure: *A DNA sequence motif represented as a sequence logo for the LexA-binding motif.*

An Example is the N-Glycosylation Site Motif: *Asn, followed by anything but Pro, followed by either Ser or Thr, followed by anything but Pro* where the three-letter abbreviations are the conventional designations for amino acids.

Overview

When a sequence motif appears in the exon of a gene, it may encode the "structural motif" of a protein, that is a stereotypical element of the overall structure of the protein. Nevertheless, motifs need not be associated with a distinctive secondary structure. "Noncoding" sequences are not translated into proteins, and nucleic acids with such motifs need not deviate from the typical shape (e.g. the "B-form" DNA double helix).

Outside of gene exons, there exist regulatory sequence motifs and motifs within the "junk", such as satellite DNA. Some of these are believed to affect the shape of nucleic acids, but this is only sometimes the case. For example, many DNA binding proteins that have affinity for specific DNA binding sites bind DNA in only its double-helical form. They are able to recognise motifs through contact with the double helix's major or minor groove.

Short coding motifs, which appear to lack secondary structure, include those that label proteins for delivery to particular parts of a cell, or mark them for phosphorylation. Within a sequence or database of sequences, researchers search and find motifs using computer-based techniques of sequence analysis, such as BLAST. Such techniques belong to the discipline of bioinformatics.

Motif Representation

Consider the N-Glycosylation Site Motif Mentioned above: *Asn, followed by anything but Pro, followed by either Ser or Thr, followed by anything but Pro*

This pattern may be written as N{P}[ST]{P} where N = Asn, P = Pro, S = Ser, T = Thr; {X} means any amino acid except X; and [XY] means either X or Y.

The notation [XY] does not give any indication of the probability of X or Y occurring in the pattern. Observed probabilities can be graphically represented using sequence logos. Sometimes patterns are defined in terms of a probabilistic model such as a hidden Markov model.

Motifs and Consensus Sequences

The notation [XYZ] means X or Y or Z, but does not indicate the likelihood of any particular match. For this reason, two or more patterns are often associated with a single motif: the defining pattern, and various typical patterns.

For example, the defining sequence for the IQ motif may be taken to be:

```
[FILV]Qxxx[RK]Gxxx[RK]xx[FILVWY]
```

where x signifies any amino acid, and the square brackets indicate an alternative.

Usually, however, the first letter is I, and both [RK] choices resolve to R. Since the last choice is so wide, the pattern IQxxxRGxxxR is sometimes equated with the IQ motif itself, but a more accurate description would be a *consensus sequence for the IQ motif.*

De Novo Computational Discovery of Motifs

There are software programs which, given multiple input sequences, attempt to identify one or more candidate motifs. One example is MEME, which generates statistical information for each candidate. A related algorithm, EXTREME, can discover motifs thousands to millions of times faster than MEME. Other algorithms include AlignAce, Amadeus, CisModule, FIRE, Gibbs Motif Sampler, PhyloGibbs, SeSiMCMC, ChIPMunk and Weeder. SCOPE, MotifVoter, and MProfiler are ensemble motif finders that uses several algorithms simultaneously. The planted motif search is another motif discovery method that is based on combinatorial approach. There currently exist more than 100 publications with similar algorithms; Weirauch *et al.* evaluated many related algorithms in a 2013 benchmark.

Discovery through Evolutionary Conservation

Motifs have been discovered by studying similar genes in different species. For example, by aligning the amino acid sequences specified by the GCM (*glial cells missing*) gene in man, mouse and *D. melanogaster*,

Akiyama and others discovered a pattern which they called the GCM motif. It spans about 150 amino acid residues, and begins as follows:

WDIND*.*P..*...D.F.*W***.**.IYS**...A.*H*S*WAMRNTNNHN

Here each . signifies a single amino acid or a gap, and each * indicates one member of a closely related family of amino acids.

The authors were able to show that the motif has DNA binding activity. PhyloGibbs and the Gibbs Motif Sampler are motif discovery algorithms that consider phylogenetic conservation.

Pattern Description Notations

Several notations for describing motifs are in use but most of them are variants of standard notations for regular expressions and use these conventions:

- there is an alphabet of single characters, each denoting a specific amino acid or a set of amino acids;
- a string of characters drawn from the alphabet denotes a sequence of the corresponding amino acids;
- any string of characters drawn from the alphabet enclosed in square brackets matches any one of the corresponding amino acids; e.g. [abc] matches any of the amino acids represented by a or b or c.

The fundamental idea behind all these notations is the matching principle, which assigns a meaning to a sequence of elements of the pattern notation:

> *a sequence of elements of the pattern notation matches a sequence of amino acids if and only if the latter sequence can be partitioned into subsequences in such a way that each pattern element matches the corresponding subsequence in turn.*

Thus the pattern [AB] [CDE] F matches the six amino acid sequences corresponding to ACF, ADF, AEF, BCF, BDF, and BEF.

Different pattern description notations have other ways of forming pattern elements. One of these notations is the PROSITE notation, described in the following subsection.

PROSITE Pattern Notation

The PROSITE notation uses the IUPAC one-letter codes and conforms to the above description with the exception that a concatenation symbol, '-', is used between pattern elements, but it is often dropped between letters of the pattern alphabet.

PROSITE allows the following pattern elements in addition to those described previously:

- The lower case letter ‘x’ can be used as a pattern element to denote any amino acid.
- A string of characters drawn from the alphabet and enclosed in braces (curly brackets) denotes any amino acid except for those in the string. For example, {ST} denotes any amino acid other than S or T.
- If a pattern is restricted to the N-terminal of a sequence, the pattern is prefixed with ‘<‘.
- If a pattern is restricted to the C-terminal of a sequence, the pattern is suffixed with ‘>’.
- The character ‘>’ can also occur inside a terminating square bracket pattern, so that S[T>] matches both “ST” and “S>”.
- If e is a pattern element, and m and n are two decimal integers with m <= n, then:
    - e(m) is equivalent to the repetition of e exactly m times;
    - e(m,n) is equivalent to the repetition of e exactly k times for any integer k satisfying: m <= k <= n.

Some examples:

- x(3) is equivalent to x-x-x.
- x(2,4) matches any sequence that matches x-x or x-x-x or x-x-x-x.

The signature of the C2H2-type *zinc finger* domain is:

- C-x(2,4)-C-x(3)-[LIVMFYWC]-x(8)-H-x(3,5)-H

Matrices

A matrix of numbers containing scores for each residue or nucleotide at each position of a fixed-length motif. There are two types of weight matrices.

- A position frequency matrix (PFM) records the position-dependent frequency of each residue or nucleotide. PFMs can be experimentally determined from SELEX experiments or computationally discovered by tools such as MEME using hidden Markov models.
- A position weight matrix (PWM) contains log odds weights for computing a match score. A cutoff is needed to specify whether an input sequence matches the motif or not. PWMs are calculated from PFMs.

An example of a PFM from the TRANSFAC database for the transcription factor AP-1:

Pos	*A*	*C*	*G*	*T*	*IUPAC*
01	6	2	8	1	R
02	3	5	9	0	S
03	0	0	0	17	T
04	0	0	17	0	G
05	17	0	0	0	A
06	0	16	0	1	C
07	3	2	3	9	T
08	4	7	2	4	N
09	9	6	1	1	M
10	4	3	7	3	N
11	6	3	1	7	W

The first column specifies the position, the second column contains the number of occurrences of A at that position, the third column contains the number of occurrences of C at that position, the fourth column contains the number of occurrences of G at that position, the fifth column contains the number of occurrences of T at that position, and the last column contains the IUPAC notation for that position. Note that the sums of occurrences for A, C, G, and T for each row should be equal because the PFM is derived from aggregating several consensus sequences.

Chapter 2

Comparative Genomics

Comparative genomics is a field of biological research in which the genomic features of different organisms are compared. The genomic features may include the DNA sequence, genes, gene order, regulatory sequences, and other genomic structural landmarks. In this branch of genomics, whole or large parts of genomes resulting from genome projects are compared to study basic biological similarities and differences as well as evolutionary relationships between organisms.

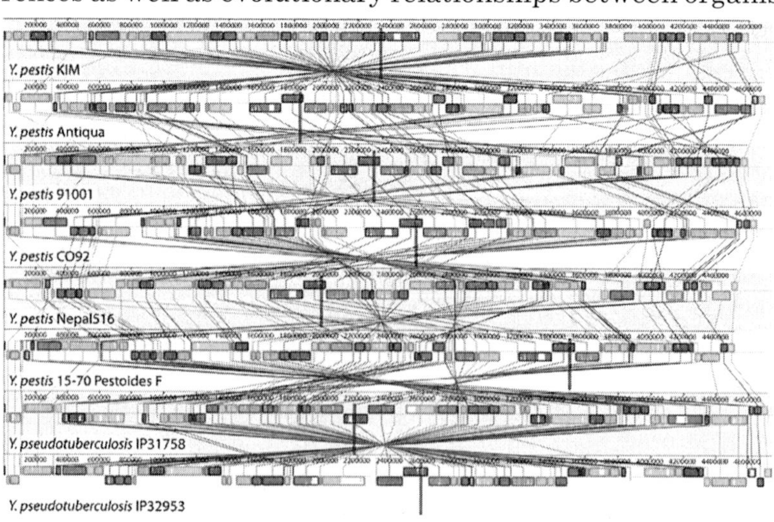

Figure: *Whole genome alignment is a typical method in comparative genomics. This alignment of eight* Yersinia *bacteria genomes reveals 78 locally collinear blocks conserved among all eight taxa. Each chromosome has been laid out horizontally and homologous blocks in each genome are shown as identically coloured regions linked across genomes. Regions that are inverted relative to* Y. pestis *KIM are shifted below a genome's centre axis.*

The major principle of comparative genomics is that common features of two organisms will often be encoded within the DNA that is evolutionary conserved between them. Therefore, comparative genomic approaches start with making some form of alignment of genome sequences and looking for orthologous sequences (sequences that share a common ancestry) in the aligned genomes and checking to what extent those sequences are conserved. Based on these, inference on genome and molecular evolution are made and this may in turn be put in the context of, for example, phenotypic evolution or population genetics.

Virtually started as soon as the whole genomes of two organisms became available (that is, the genomes of the bacteria *Haemophilus influenzae* and *Mycoplasma genitalium*) in 1995, comparative genomics is now a standard component of the analysis of every new genome sequence. With the explosion in the number of genome project due to the advancements in DNA sequencing technologies, particularly the next-generation sequencing methods in late 2000s, this field has become more sophisticated, making it possible to deal with many genomes in a single study. Comparative genomics has revealed high levels of similarity between closely related oganisms, such as humans and chimpanzees, and, more surprisingly, similarity between seemingly distantly related organisms, such as humans and the yeast *Saccharomyces cerevisiae*. It has also showed the extreme diversity of the gene composition in different evolutionary lineages.

History

Comparative genomics has a root in the comparison of virus genomes in the early 1980s. For example, small RNA viruses infecting animals (picornaviruses) and those infecting plants (cowpea mosaic virus) were compared and turned out to share significant sequence similarity and, in part, the order of their genes. In 1986, the first comparative genomic study at a larger scale was published, comparing the genomes of varicella-zoster virus and Epstein-Barr virus that contained more than 100 genes each.

The first complete genome sequence of a cellular organism, that of *Haemophilus influenzae* Rd, was published in 1995. The second genome sequencing paper was of the small parasitic bacterium *Mycoplasma genitalium* published in the same year. Starting from this paper, reports on new genomes inevitably became comparative-genomic studies.

Saccharomyces cerevisiae, the baker's yeast, was the first eukaryote to have its complete genome sequence published in 1996. After the publication of the roundworm *Caenorhabditis elegans* genome in 1998 and together with the fruit fly *Drosophila melanogaster* genome in 2000, Gerald M. Rubin and his team published a paper titled "Comparative Genomics of the Eukaryotes", in which they compared the genomes of the eukaryotes *D. melanogaster*, *C. elegans*, and *S. cerevisiae*, as well as the prokaryote *H. influenzae*.

With the publication of the large genomes of vertebrates in the 2000s, including human, the Japanese pufferfish *Takifugu rubripes*, and mouse, precomputed results of large genome comparisons have been released for downloading or for visualization in a genome browser. Instead of undertaking their own analyses, most biologists can access these large cross-species comparisons and avoid the impracticality caused by the size of the genomes.

Next-generation sequencing methods, which were first introduced in 2007, have produced an enormous amount of genomic data and have allowed researchers to generate multiple (prokaryotic) draft genome sequences at once. These methods can also quickly uncover single-nucleotide polymorphisms, insertions and deletions by mapping unassembled reads against a well annotated reference genome, and thus provide a list of possible gene differences that may be the basis for any functional variation among strains.

Evolutionary Principles

One character of Biology is evolution, evolutionary theory is also the theoretical foundation of comparative genomics, and at the same time the results of comparative genomics unprecedentedly enriched and developed the theory of evolution. When two or more of the genome sequence compared, in essence get the evolutionary relationships of the sequence in the phylogenetic tree. Increased genome information of study makes molecular evolution, gene function at the genome level possible. Based on a variety of biological genome data and the study of vertical and horizontal evolution process, can understand vital parts of gene structure and its regulation function for life. But as a result in biological genome about 1.5% ~ 14.5% of the genes related to "lateral migration phenomenon", namely the gene transfer between populations which can exist at the same time, the differences in sequence result has nothing to do with the evolution. So in the system analysis, it needs to establish a relatively complete evolution model, in order to avoid gene transfer and the influence of the lack of more appropriate species which are conserved sequence.

Similarity of related genomes is the basis of comparative genomics. Two creatures which have a recent common ancestor, the species difference genomes between them were evolved from ancestors' genome, the closer the two organisms on the evolutionary stages, the higher their genome correlated. If there is close relationship between them, then their genome will behave like linear (synteny), namely some or all of the genetic sequences are conservative. So scientists can use the homology of the sequence and structure of encoding between mode genomes, by known genome mapping information to locate other genes in the genome, so as to reveal the potential function of the genes, clarify evolutionary relationship and the inner structure of the genome.

Orthologous sequences are separate because of speciation: a gene exists in the original species, the species divided into two species, so genes in new species are orthologous. Paralogy sequences are separate by gene cloning (gene duplication): if a particular gene in the biology is copied, then the copy of the two sequences is paralogy. A pair of orthologous sequences is called orthologous pairs (orthologs), a pair of paralogy sequence is called collateral pairs (paralogs). Orthologous pairs usually have the same or similar function, but not necessarily on collateral pairs: due to the lack of the power of natural selection, the original duplicate copy of the genes are variation and get free new functions.

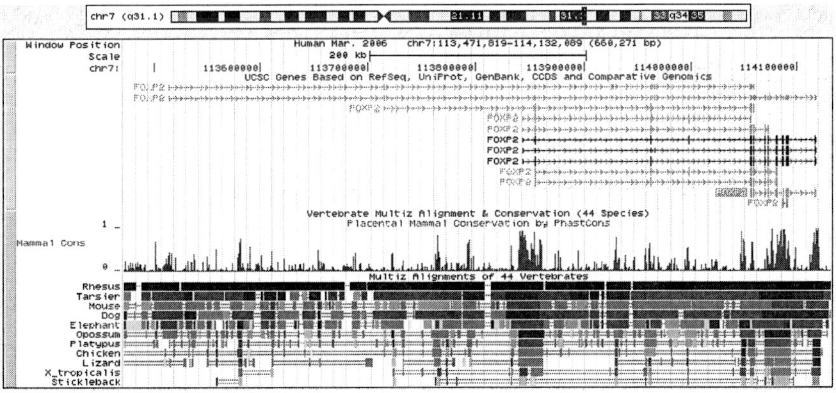

Figure: *Human FOXP2 gene and evolutionary conservation is shown in and multiple alignment (at bottom of figure) in this image from the UCSC Genome Browser. Note that conservation tends to cluster around coding regions (exons).*

Comparative genomics exploits both similarities and differences in the proteins, RNA, and regulatory regions of different organisms to infer how selection has acted upon these elements. Those elements that are responsible for similarities between different species should be conserved through time (stabilizing selection), while those elements responsible for differences among species should be divergent (positive selection). Finally, those elements that are unimportant to the evolutionary success of the organism will be unconserved (selection is neutral).

One of the important goals of the field is the identification of the mechanisms of eukaryotic genome evolution. It is however often complicated by the multiplicity of events that have taken place throughout the history of individual lineages, leaving only distorted and superimposed traces in the genome of each living organism. For this reason comparative genomics studies of small model organisms (for example the model Caenorhabditis elegans and closely related Caenorhabditis briggsae) are of great importance to advance our understanding of general mechanisms of evolution.

Methods

Computational approaches to genome comparison have recently become a common research topic in computer science. A public collection of case studies and demonstrations is growing, ranging from whole genome comparisons to gene expression analysis. This has increased the introduction of different ideas, including concepts from systems and control, information theory, strings analysis and data mining. It is anticipated that computational approaches will become and remain a standard topic for research and teaching, while multiple courses will begin training students to be fluent in both topics.

Tools

Computational tools for analyzing sequences and complete genomes are developed quickly due to the availability of large amount of genomic data. At the same time, comparative analysis tools are progressed and improved. In the challenges about these analyses, it is very important to visualize the comparative results.

Visualization of sequence conservation is a tough task of comparative sequence analysis. As we know, it is highly inefficient to examine the alignment of long genomic regions manually. Internet-based genome browsers provide many useful tools for investigating genomic sequences due to integrating all sequence-based biological information on genomic regions. When we extract large amount of relevant biological data, they can be very easy to use and less time-consuming.

- UCSC Browser: This site contains the reference sequence and working draft assemblies for a large collection of genomes.
- Ensembl: The Ensembl project produces genome databases for vertebrates and other eukaryotic species, and makes this information freely available online.

- MapView: The Map Viewer provides a wide variety of genome mapping and sequencing data.
- VISTA is a comprehensive suite of programs and databases for comparative analysis of genomic sequences. It was built to visualize the results of comparative analysis based on DNA alignments. The presentation of comparative data generated by VISTA can easily suit both small and large scale of data.

An advantage of using online tools is that these websites are being developed and updated constantly. There are many new settings and content can be used online to improve efficiency.

Gene Expression Profiling

In the field of molecular biology, gene expression profiling is the measurement of the activity (the expression) of thousands of genes at once, to create a global picture of cellular function. These profiles can, for example, distinguish between cells that are actively dividing, or show how the cells react to a particular treatment. Many experiments of this sort measure an entire genome simultaneously, that is, every gene present in a particular cell.

DNA microarray technology measures the relative activity of previously identified target genes. Sequence based techniques, like serial analysis of gene expression (SAGE, SuperSAGE) are also used for gene expression profiling. SuperSAGE is especially accurate and can measure any active gene, not just a predefined set. The advent of next-generation sequencing has made sequence based expression analysis an increasingly popular, "digital" alternative to microarrays called RNA-Seq. However, microarrays are far more common, accounting for 17,000 PubMed articles by 2006.

Background

Expression profiling is a logical next step after sequencing a genome: the sequence tells us what the cell could possibly do, while the expression profile tells us what it is actually doing at a point in time. Genes contain the instructions for making messenger RNA (mRNA), but at any moment each cell makes mRNA from only a fraction of the genes it carries. If a gene is used to produce mRNA, it is considered "on", otherwise "off". Many factors determine whether a gene is on or off, such as the time of day, whether or not the cell is actively dividing, its local environment, and chemical signals from other cells. For instance, skin cells, liver cells and nerve cells turn on (express) somewhat different genes and that is in large part what makes them different. Therefore, an expression profile allows one to deduce a cell's type, state, environment, and so forth.

Expression profiling experiments often involve measuring the relative amount of mRNA expressed in two or more experimental conditions. This is because altered levels of a specific sequence of mRNA suggest a changed need for the protein coded by the mRNA, perhaps indicating a homeostatic response or a pathological condition. For example, higher levels of mRNA coding for alcohol dehydrogenase suggest that the cells or tissues under study are responding to increased levels of ethanol in their environment. Similarly, if breast cancer cells express higher levels of mRNA associated with a particular transmembrane receptor than normal cells do, it might be that this receptor plays a role in breast cancer. A drug that interferes with this receptor may prevent or treat breast cancer. In developing a drug, one may perform gene expression profiling experiments to help assess the drug's toxicity, perhaps by looking for changing levels in the expression of cytochrome P450 genes, which may be a biomarker of drug metabolism. Gene expression profiling may become an important diagnostic test.

Comparison to Proteomics

The human genome contains on the order of 25,000 genes which work in concert to produce on the order of 1,000,000 distinct proteins. This is due to alternative splicing, and also because cells make important changes to proteins through post-translational modification after they first construct them, so a given gene serves as the basis for many possible versions of a particular protein. In any case, a single mass spectrometry experiment can identify about 2,000 proteins or 0.2% of the total. While knowledge of the precise proteins a cell makes (proteomics) is more relevant than knowing how much messenger RNA is made from each gene, gene expression profiling provides the most global picture possible in a single experiment.

Use in Hypothesis Generation and Testing

Sometimes, a scientist already has an idea what is going on, a hypothesis, and he or she performs an expression profiling experiment with the idea of potentially disproving this hypothesis. In other words, the scientist is making a specific prediction about levels of expression that could turn out to be false.

More commonly, expression profiling takes place before enough is known about how genes interact with experimental conditions for a testable hypothesis to exist. With no hypothesis, there is nothing to disprove, but expression profiling can help to identify a candidate hypothesis for future experiments. Most early expression profiling experiments, and many current ones, have this form which is known as class discovery. A popular approach to class discovery involves grouping similar genes or samples together using k-means or hierarchical clustering. Apart from selecting a clustering algorithm, user usually has to choose an appropriate proximity measure (distance or similarity) between data objects. The figure above represents the output of a two dimensional cluster, in which similar samples (rows, above) and similar gene probes (columns) were organised so that they would lie close together. The simplest form of class discovery would be to list all the genes that changed by more than a certain amount between two experimental conditions.

Class prediction is more difficult than class discovery, but it allows one to answer questions of direct clinical significance such as, given this profile, what is the probability that this patient will respond to this drug? This requires many examples of profiles that responded and did not respond, as well as cross-validation techniques to discriminate between them.

Limitations

In general, expression profiling studies report those genes that showed statistically significant differences under changed experimental conditions. This is typically a small fraction of the genome for several reasons. First, different cells and tissues express a subset of genes as a direct consequence of cellular differentiation so many genes are turned off. Second, many of the genes code for proteins that are required for survival in very specific amounts so many genes do not change. Third, cells use many other mechanisms to regulate proteins in addition to altering the amount of mRNA, so these genes may stay consistently expressed even when protein concentrations are rising and falling. Fourth, financial constraints limit expression profiling experiments to a small number of observations of the same gene under identical conditions, reducing the statistical power of the experiment, making it impossible for the experiment to identify important but subtle changes. Finally, it takes a great amount of effort to discuss the biological significance of each regulated gene, so scientists often limit their discussion to a subset. Newer microarray analysis techniques automate certain aspects of attaching biological significance to expression profiling results, but this remains a very difficult problem.

The relatively short length of gene lists published from expression profiling experiments limits the extent to which experiments performed in different laboratories appear to agree. Placing expression profiling results in a publicly accessible microarray database makes it possible for researchers to assess expression patterns beyond the scope of published results, perhaps identifying similarity with their own work.

Validation of High throughput Measurements

Both DNA microarrays and quantitative PCR exploit the preferential binding or "base pairing" of complementary nucleic acid sequences, and both are used in gene expression profiling, often in a serial fashion. While high throughput DNA microarrays lack the quantitative accuracy of qPCR, it takes about the same time to measure the gene expression of a few dozen genes via qPCR as it would to measure an entire genome using DNA microarrays. So it often makes sense to perform semi-quantitative DNA microarray analysis experiments to identify candidate genes, then perform qPCR on some of the most interesting candidate genes to validate the microarray results. Other experiments, such as a Western blot of some of the protein products of differentially expressed genes, make conclusions based on the expression profile more persuasive, since the mRNA levels do not necessarily correlate to the amount of expressed protein.

Statistical Analysis

Data analysis of microarrays has become an area of intense research. Simply stating that a group of genes were regulated by at least twofold, once a common practice, lacks a solid statistical footing. With five or fewer replicates in each group, typical for microarrays, a single outlier observation can create an apparent difference greater than two-fold. In addition, arbitrarily setting the bar at two-fold is not biologically sound, as it eliminates from consideration many genes with obvious biological significance.

Rather than identify differentially expressed genes using a fold change cutoff, one can use a variety of statistical tests or omnibus tests such as ANOVA, all of which consider both fold change and variability to create a p-value, an estimate of how often we would observe the data by chance alone. Applying p-values to microarrays is complicated by the large number of multiple comparisons (genes) involved. For example, a p-value of 0.05 is typically thought to indicate significance, since it estimates a 5% probability of observing the data by chance. But with 10,000 genes on a microarray, 500 genes would be identified as significant at $p < 0.05$ even if there were no difference between the experimental groups. One obvious solution is to consider significant only those genes meeting a much more stringent p value criterion, e.g., one could perform a Bonferroni correction on the p-values, or use a false discovery rate calculation to adjust p-values in proportion to the number of parallel tests involved. Unfortunately, these approaches may reduce the number of significant genes to zero, even when genes are in fact differentially expressed. Current statistics such as Rank products aim to strike a balance between false discovery of genes due to chance variation and non-discovery of differentially expressed genes. Commonly cited methods include the Significance Analysis of Microarrays (SAM) and a wide variety of methods are available from Bioconductor and a variety of analysis packages from bioinformatics companies.

Selecting a different test usually identifies a different list of significant genes since each test operates under a specific set of assumptions, and places a different emphasis on certain features in the data. Many tests begin with the assumption of a normal distribution in the data, because that seems like a sensible starting point and often produces results that appear more significant. Some tests consider the joint distribution of all gene observations to estimate general variability in measurements, while others look at each gene in isolation. Many modern microarray analysis techniques involve bootstrapping (statistics), machine learning or Monte Carlo methods.

As the number of replicate measurements in a microarray experiment increases, various statistical approaches yield increasingly similar results, but lack of concordance between different statistical methods makes array results appear less trustworthy. The MAQC Project makes recommendations to guide researchers in selecting more standard methods (e.g. using p-value and fold-change together for selecting the differentially expressed genes) so that experiments performed in different laboratories will agree better.

Different from the analysis on differentially expressed individual genes, another type of analysis focuses on differential expression or perturbation of pre-defined gene sets and is called gene set analysis. Gene set analysis demonstrated several major advantages over individual gene differential expression analysis. Gene sets are groups of genes that are functionally related according to current knowledge. Therefore, gene set analysis is considered a knowledge based analysis approach. Commonly used gene sets include those derived from KEGG pathways, Gene Ontology terms, gene groups that share some other functional annotations, such as common transcriptional regulators etc. Representative gene set analysis methods include GSEA, which estimates significance of gene sets based on permutation of sample labels, and GAGE, which tests the significance of gene sets based on permutation of gene labels or a parametric distribution.

Gene Annotation

While the statistics may identify which gene products change under experimental conditions, making biological sense of expression profiling rests on knowing which protein each gene product makes and what function this protein performs. Gene annotation provides functional and other information, for example the location of each gene within a particular chromosome. Some functional annotations are more reliable than others; some are absent. Gene annotation databases change regularly, and various databases refer to the same protein by different names, reflecting a changing understanding of protein function. Use of standardized gene nomenclature helps address the naming aspect of the problem, but exact matching of transcripts to genes remains an important consideration.

Categorizing Regulated Genes

Having identified some set of regulated genes, the next step in expression profiling involves looking for patterns within the regulated set. Do the proteins made from these genes perform similar functions? Are they chemically similar? Do they reside in similar parts of the cell? Gene ontology analysis provides a standard way to define these relationships. Gene ontologies start with very broad categories, e.g., "metabolic process" and break them down into smaller categories, e.g., "carbohydrate metabolic process" and finally into quite restrictive categories like "inositol and derivative phosphorylation".

Genes have other attributes beside biological function, chemical properties and cellular location. One can compose sets of genes based on proximity to other genes, association with a disease, and relationships with drugs or toxins. The Molecular Signatures Database and the Comparative Toxicogenomics Database are examples of resources to categorize genes in numerous ways.

Finding Patterns among Regulated Genes

Regulated genes are categorized in terms of what they are and what they do, important relationships between genes may emerge. For example, we might see evidence that a certain gene creates a protein to make an enzyme that activates a protein to turn on a second gene on our list. This second gene may be a transcription factor that regulates yet another gene from our list. Observing these links we may begin to suspect that they represent much more than chance associations in the results, and that they are all on our list because of an underlying biological process. On the other hand, it could be that if one selected genes at random, one might find many that seem to have something in common. In this sense, we need rigorous statistical procedures to test whether the emerging biological themes is significant or not. That is where gene set analysis comes in.

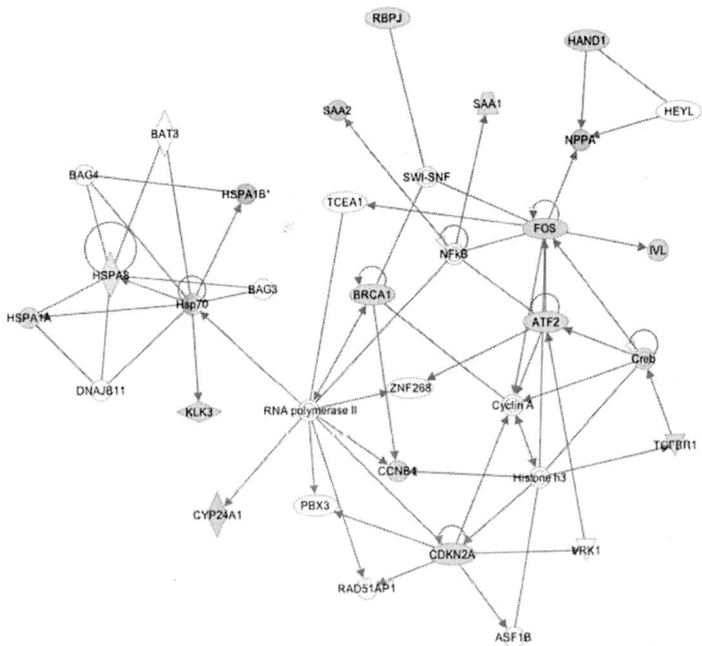

Figure: *Ingenuity Gene Network Diagram which dynamically assembles genes with known relationships. Green indicates reduced expression, red indicates increased expression. The algorithm has included unregulated genes, white, to improve connectivity.*

Cause and Effect Relationships

Fairly straightforward statistics provide estimates of whether associations between genes on lists are greater than what one would expect by chance. These statistics are interesting, even if they represent a substantial oversimplification of what is really going on. Here is an example. Suppose there are 10,000 genes in an experiment, only 50 (0.5%) of which play a known role in making cholesterol. The experiment identifies 200 regulated genes. Of those, 40 (20%) turn out to be on a list of cholesterol genes as well. Based on the overall prevalence of the cholesterol genes (0.5%) one expects an average of 1 cholesterol gene for every 200 regulated genes, that is, 0.005 times 200. This expectation is an average, so one expects to see more than one some of the time. The question becomes how often we would see 40 instead of 1 due to pure chance. According to the hypergeometric distribution, one would expect to try about 10^57 times (10 followed by 56 zeroes) before picking 39 or more of the cholesterol genes from a pool of 10,000 by drawing 200 genes at random. Whether one pays much attention to how infinitesimally small the probability of observing this by chance is, one would conclude that the regulated gene list is enriched in genes with a known cholesterol association.

One might further hypothesize that the experimental treatment regulates cholesterol, because the treatment seems to selectively regulate genes associated with cholesterol. While this may be true, there are a number of reasons why making this a firm conclusion based on enrichment alone represents an unwarranted leap of faith. One previously mentioned issue has to do with the observation that gene regulation may have no direct impact on protein regulation: even if the proteins coded for by these genes do nothing other than make cholesterol, showing that their mRNA is altered does not directly tell us what is happening at the protein level. It is quite possible that the amount of these cholesterol-related proteins remains constant under the experimental conditions. Second, even if protein levels do change, perhaps there is always enough of them around to make cholesterol as fast as it can be possibly made, that is, another protein, not on our list, is the rate determining step in the process of making cholesterol. Finally, proteins typically play many roles, so these genes may be regulated not because of their shared association with making cholesterol but because of a shared role in a completely independent process.

Bearing the foregoing caveats in mind, while gene profiles do not in themselves prove causal relationships between treatments and biological effects, they do offer unique biological insights that would often be very difficult to arrive at in other ways.

Using Patterns to Find Regulated Genes

As described above, one can identify significantly regulated genes first and then find patterns by comparing the list of significant genes to sets of genes known to share certain associations. One can also work the problem in reverse order. Here is a very simple example. Suppose there are 40 genes associated with a known process, for example, a predisposition to diabetes. Looking at two groups of expression profiles, one for mice fed a high carbohydrate diet and one for mice fed a low carbohydrate diet, one observes that all 40 diabetes genes are expressed at a higher level in the high carbohydrate group than the low carbohydrate group. Regardless of whether any of these genes would have made it to a list of significantly altered genes, observing all 40 up, and none down appears unlikely to be the result of pure chance: flipping 40 heads in a row is predicted to occur about one time in a trillion attempts using a fair coin.

For a type of cell, the group of genes whose combined expression pattern is uniquely characteristic to a given condition constitutes the gene signature of this condition. Ideally, the gene signature can be used to select a group of patients at a specific state of a disease with accuracy that facilitates selection of treatments. Gene Set Enrichment Analysis (GSEA) and similar methods take advantage of this kind of logic but uses more sophisticated statistics, because component genes in real processes display more complex behaviour than simply moving up or down as a group, and the amount the genes move up and down is meaningful, not just the direction. In any case, these statistics measure how different the behaviour of some small set of genes is compared to genes not in that small set.

GSEA uses a Kolmogorov Smirnov style statistic to see whether any previously defined gene sets exhibited unusual behaviour in the current expression profile. This leads to a multiple hypothesis testing challenge, but reasonable methods exist to address it.

Conclusions

Expression profiling provides new information about what genes do under various conditions. Overall, microarray technology produces reliable expression profiles. From this information one can generate new hypotheses about biology or test existing ones. However, the size and complexity of these experiments often results in a wide variety of possible interpretations. In many cases, analyzing expression profiling results takes far more effort than performing the initial experiments.

Most researchers use multiple statistical methods and exploratory data analysis before publishing their expression profiling results, coordinating their efforts with a bioinformatician or other expert in microarray technology. Good experimental design, adequate biological replication and follow up experiments play key roles in successful expression profiling experiments.

Gene Expression Profiling in Cancer

Cancer is a disease characterized by uncontrolled cell growth and proliferation. For cancer to develop, genes regulating cell growth and differentiation must be altered; these mutations are then maintained through subsequent cell divisions and are thus present in all cancerous cells. Gene expression profiling is a technique used in molecular biology to query the expression of thousands of genes simultaneously. In the context of cancer, gene expression profiling has been used to more accurately classify tumors. The information derived from gene expression profiling often has an impact on predicting the patient's clinical outcome.

Background

Oncogenesis: Oncogenesis is the process by which normal cells acquire the properties of cancer cells leading to the formation of a cancer or tumor. It is characterized by a molecular reprogramming of a cell to undergo uninhibited cell division, allowing the formation of a malignant mass. The cells forming this mass undergo natural selection: as cells acquire mutations that enhance their survivability or reproductive capacity, they dominate the growing tumor as other cells are out-competed. Because of these selective properties, the majority of cells within a tumor will share a common profile of gene expression.

Gene Expression Profiling

Gene expression profiling is a technique used in molecular biology to query the expression of thousands of genes simultaneously. While almost all cells in an organism contain the entire genome of the organism, only a small subset of those genes is expressed as messenger RNA (mRNA) at any given time, and their relative expression can be evaluated. Techniques include DNA microarray technology or sequenced-based techniques such as serial analysis of gene expression (SAGE).

Current cancer research makes use primarily of DNA microarrays in which an arrayed series of microscopic spots of pre-defined DNA oligonucleotides known as *probes* are covalently attached to a solid surface such as glass, forming what is known as a gene chip. DNA labelled with fluorophores (*target*) is prepared from a sample such as a tumor biopsy and is hybridized to the complementary DNA (cDNA) sequences on the gene chip. The chip is then scanned for the presence and strength of the fluorescent labels at each spot representing probe-target hybrids. The level of fluorescence at a particular spot provides quantitative information about the expression of the particular gene corresponding to the spotted cDNA sequence. DNA microarrays evolved from Southern blotting which allows for detection of a specific DNA sequence in a sample of DNA.

Due to lowering costs, RNA-Sequencing is becoming more common as a method for cancer gene expression profiling. It is superior to microarray techniques due to not having the bias inherent in probe selection.

Cancer Classification

Background

Classification of cancers has been dominated by the fields of histology and histopathology which aim to leverage morphological markers for accurate identification of a tumor type. Histological methods rely on chemical staining of tissues with pigments such as haematoxylin and eosin and microscopy-based visualization by a pathologist. The identification of tumor subtypes is based on established classification schemes such as the International Classification of Diseases published by the World Health Organisation which provides codes to classify diseases and a wide variety of signs, symptoms, abnormal findings, complaints, social circumstances, and external causes of injury or diseases. For some types of cancer, these methods are unable to distinguish between subclasses; for example, defining subgroups of diffuse large B-cell lymphoma (DLBCL) have largely failed due to discrepancies between inter- and intra-observer reproducibility. Furthermore, the clinical outcomes of tumors classified as DLBCLs is highly variable suggesting that there are multiple subtypes of DLBCL that cannot be distinguished based on these histological markers. Breast tumor classification too has largely failed based on these predictors. Development of effective therapies depends on accurate diagnosis; additionally, poor diagnosis can lead to patient suffering due to needless side-effects from non-targeted treatments and to increased health care expenditure. Most telling perhaps is that 70-80% of breast cancer patients receiving chemotherapy based on traditional predictors would have survived without it.

Of note, similar gene expression patterns associated with metastatic behaviour of breast cancer tumor cells have also been found in breast cancer of dog, the most common tumor of the female dog.

Presented below are ways that gene expression profiling has been used to more precisely classify tumors into subgroups, often with clinical impact.

Molecular Signature

In a particular type of cell or tissue, only a small subset of an organism's genomic DNA will be expressed as mRNAs at any given time. The unique pattern of gene expression for a given cell or tissue is referred to as its molecular signature. For example, the expression of genes in skin cells would be very different compared to those expressed in blood cells. Microarray analysis can provide quantitative gene expression information allowing for the generation of a molecular signature, each unique to a particular class of tumor. This idea was first shown experimentally in 2000 by researchers at Stanford University published in *Nature Genetics*. The authors measured the relative expression of 9,703 human cDNAs in sixty cancer cell lines previously studied and characterized by the National Cancer Institute's Developmental Therapeutics Program. A hierarchical clustering algorithm was used to group cell lines based on the similarity by which the pattern of gene expression varied. In this study by Ross *et al.*, the majority of cell lines with common organs of origin (based on information from the National Institutes of Health) clustered together at terminal branches, suggesting that cancer cells arising from the same tissue share many molecular characteristics. This allows for reliable identification of tumor type based on gene expression.

Tumor Subclass

A more powerful result of gene expression profiling is the ability to further classify tumors into subtypes having distinct biological properties and impact on prognoses. For example, some diffuse large B-cell lymphomas (DLBCLs) are indistinguishable based on histological methods yet are clinically heterogeneous: 40% of patients respond well and exhibit prolonged survival while the remaining 60% do not.

In 2000, Stanford researchers published results in *Nature*, utilising expression profiling techniques to stratify DLBCL to two subtypes: germinal centre B-like DLBCL and activated B-like DLBCL. The authors developed custom microarrays termed "lymphochips" that were used to query expression of 17,856 genes preferentially expressed in lymphoid cells and those with roles in cancer or immunology for 96 lymphocyte samples. The hierarchical clustering algorithm identified a subset of tumors that would have been labelled DLBCLs by traditional histological methods; however, the expression profiles of these tumors were heterogeneous.

When the tumors were re-clustered based on expression of germinal centre B-cell genes, a second group of genes characteristic of activated B-cells emerged and were oppositely regulated compared to the first set of genes. Based on these expression patterns, the heterogeneous DLBCL cluster was subclassified to the germinal centre B-like DLBCL and activated B-like DLBCL. The distinction of these groups is significant in terms of patient overall survival: the probability of survival for patients with germinal centre B-like DLBCLs over 10 years was about 80% while those with activated B-like DLBCLs was lowered to about 40% over a shorter eight-year period.

Breast cancers are also difficult to distinguish based on histological markers. In a 2000 study published in *Nature*, Stanford researchers led by Perou, C.M. characterized gene expression patterns across 8,102 genes for 65 biopsies obtained from breast cancers. The goal of the study was to identify patterns of gene expression that could be used to describe the phenotypic diversity of breast tumors by comparing the profiles of the biopsies to those of cultured cell lines and relating this information to clinical data. The tumors were clustered into two major groups that largely reflected the ER-positive and ER-negative clinical descriptions. The ER-positive tumors were characterized by high expression of genes normally expressed in breast luminal cells. The authors suggest that this higher-order distinction may encompass at least two biologically distinct types of cancer that may each require a unique course of treatment. Within the ER-negative group, additional clusters were identified based on expression of Erb-B2 and keratin 5- and 17-enriched basal epithelial-like genes. These groups reflect distinct molecular features as related to mammary epithelial biology, based on the outcome of disease.

Clinical Application

In a 2001 study published in the *Proceedings of the National Academy of Sciences*, Sørlie *et al.* further stratified the classifications described by Perou *et al.* and explored the clinical value of these breast cancer subtypes. The authors separated the ER-positive tumors into two distinct groups and found that tumor classification based on gene expression was related to patient survival. The expression of 427 genes was measured for 78 cancers and seven non-malignant breast samples. Following hierarchical clustering, the samples formed two groups at the highest level of organisation reflecting the ER-positive and ER-negative phenotypes; the ER-negative cluster further stratified to groups identical to those described by Perou *et al.* In contrast to previous results, Sørlie *et al.* found that the ER-positive group could also be separated into three distinct subgroups termed luminal subtypes A, B, and, C based on patterns of luminal-specific gene expression with different outcomes. The authors further found once they performed survival analyses that tumors belonging to the various groups showed significantly different outcomes when treated uniformly. Survival analyses are often shown as Kaplan-Meier survival plots, an example of which is shown to the right.

In addition to identifying genes that correlate to survival, microarray analyses have been utilised to establish gene expression profiles associated with prognosis. It is agreed upon that patients with tumors exhibiting poor prognostic features would benefit the most from adjuvant therapy as these treatments substantially improve overall survival for women with breast cancer. Traditional prognostic factors, however, are inexact as mentioned above.

Researchers at the Netherlands Cancer Institute were able to identify "good-prognosis" and "bad-prognosis" signatures based on the expression of 70 genes that was better able to predict the likelihood of metastasis development within five years for breast cancer patients Metastasis involves the spread of cancer from one organ to others throughout the body and is the principal cause of death in cancer patients. While the study at the Netherlands Cancer Institute applied to breast cancer patients only, researchers at Massachusetts Institute of Technology identified a molecular signature of metastasis that applied to adenocarcinomas in general.

Interaction Network

Interaction network is a network of nodes that are connected by features. If the feature is a physical and molecular, the interaction network is molecular interactions usually found in cells. Interaction network has become a research topic in biology in recent years due to rapid progress in high throughput data production.

Protein–Protein Interaction

Protein–protein interactions (PPIs) refer to intentional physical contacts established between two or more proteins as a result of biochemical events and/or electrostatic forces.

In fact, proteins are vital macromolecules, at both cellular and systemic levels, but they rarely act alone. Diverse essential molecular processes within a cell are carried out by molecular machines that are built from a large number of protein components organised by their PPIs. Indeed, these interactions are at the core of the entire interatomics system of any living cell and so, unsurprisingly, aberrant PPIs are on the basis of multiple diseases, such as Creutzfeld-Jacob, Alzheimer's disease, and cancer.

PPIs have been studied from different perspectives: biochemistry, quantum chemistry, molecular dynamics, signal transduction, among others. All this information enables the creation of large protein interaction networks – similar to metabolic or genetic/epigenetic networks – that empower the current knowledge on biochemical cascades and disease pathogenesis, as well as provide putative new therapeutic targets.

Examples of Protein–Protein Interactions

- Signal transduction: The activity of the cell is regulated by extracellular signals. Signals propagation to inside and/or along the interior of cells depends on PPIs between the various signaling molecules. This process, called signal transduction, plays a fundamental role in many biological processes and in many diseases (e.g. Parkinson's disease and cancer).
- Transport across membranes: A protein may be carrying another protein (for example, from cytoplasm to nucleus or vice versa in the case of the nuclear pore importins).
- Cell metabolism: In many biosynthetic processes enzymes interact with each other to produce small compounds or other macromolecules.
- Muscle contraction: Physiology of muscle contraction involves several interactions. Myosin filaments act as molecular motors and by binding to actin enables filament sliding. Furthermore, members of the skeletal muscle lipid droplet-associated proteins family associate with other proteins, as activator of adipose triglyceride lipase and its coactivator comparative gene identification-58, to regulate lipolysis in skeletal muscle.

Types of Protein–Protein Interactions

Protein complex assembly can result in the formation of homo-oligomeric or hetero-oligomeric complexes. In addition to the conventional complexes, as enzyme-inhibitor and antibody-antigen, interactions can also be established between domain-domain and domain-peptide. Moreover, interactions can be classified into stable or transient, and also according to the nature of the chemical bonds established between proteins.

Homo-Oligomers vs. Hetero-Oligomers

Homo-oligomers are macromolecular complexes constituted by only one type of protein subunit. Protein subunits assembly is guided by the establishment of non-covalent interactions in the quaternary structure of the protein. Disruption of homo-oligomers in order to return to the initial individual monomers often requires denaturation of the complex. Several enzymes, carrier proteins, scaffolding proteins, and transcriptional regulatory factors carry out their functions as homo-oligomers. Distinct protein subunits interact in hetero-oligomers, which are essential to control several cellular functions. The importance of the communication between heterologous proteins is even more evident during cell signaling events and such interactions are only possible due to structural domains within the proteins (as described below).

Stable Interactions vs. Transient Interactions

Stable interactions involve proteins that interact for a long time, taking part of permanent complexes as subunits, in order to carry out structural or functional roles. These are usually the case of homo-oligomers (e.g. cytochrome c), and some hetero-oligomeric proteins, as the subunits of ATPase. On the other hand, a protein may interact briefly and in a reversible manner with other proteins in only certain cellular contexts – cell type, cell cycle stage, external factors, presence of other binding proteins, etc. – as it happens with most of the proteins involved in biochemical cascades. These are called transient interactions. For example, some G protein-coupled receptors only transiently bind to $G_{i/o}$ proteins when they are activated by extracellular ligands, while some G_q-coupled receptors, such as muscarinic receptor M3, pre-couple with G_q proteins prior to the receptor-ligand binding.

Covalent vs. Non-Covalent

Covalent interactions are those with the strongest association and are formed by disulphide bonds or electron sharing. Although being rare, these interactions are determinant in some post-translational modifications, as ubiquitination and SUMOylation. Non-covalent bonds are usually established during transient interactions by the combination of weaker bonds, such as hydrogen bonds, ionic interactions, Van der Waals forces, or hydrophobic bonds.

Techniques to Study the Molecular Structure of Protein Complexes

The molecular structures of many protein complexes have been unlocked by the technique of X-ray crystallography. The first structure to be solved by this method was that of sperm whale myoglobin by Sir John Cowdery Kendrew. In this technique the angles and intensities of a beam of X-rays diffracted by crystalline atoms are detected in a film, thus producing a three-dimensional picture of the density of electrons within the crystal.

Later, nuclear magnetic resonance also started to be applied with the aim of unravelling the molecular structure of protein complexes. One of the first examples was the structure of calmodulin-binding domains bound to calmodulin. This technique is based on the study of magnetic properties of atomic nuclei, thus determining physical and chemical properties of the correspondent atoms or the molecules. Nuclear magnetic resonance is advantageous for characterizing weak PPIs.

Properties of Protein–Protein Interface

The study of the molecular structure can give fine details about the interface that enables the interaction between proteins. When characterizing PPI interfaces it is important to take into account the type of complex. Parameters evaluated include size (measured in absolute dimensions Å^2 or in solvent-accessible surface area (SASA)), shape, complementarity between surfaces, residue interface propensities, hydrophobicity, segmentation and secondary structure, and conformational changes on complex formation.

The great majority of PPI interfaces reflects the composition of protein surfaces, rather than the protein cores, in spite of being frequently enriched in hydrophobic residues, particularly in aromatic residues. PPI interfaces are dynamic and frequently planar, although they can be globular and protruding as well. Based on three structures – insulin dimer, trypsin-pancreatic trypsin inhibitor complex, and oxyhaemoglobin – Cyrus Chothia and Joel Janin found that between 1,130 and 1,720 Å of surface area was removed from contact with water indicating that hydrophobicity is a major factor of stabilization of PPIs. Later studies refined the buried surface area of the majority of interactions to 1,600±350 Å^2. However, much larger interaction interfaces were also observed and were associated with significant changes in conformation of one of the interaction partners. PPIs interfaces exhibit both shape and electrostatic complementarity.

It has been shown that PPI interfaces assemble to mitigate the protein–water interfacial tension (epistructural tension) of individual binding partners. The epistructural tension is mostly caused by a type of protein structural defect known as dehydron.

Factors that Regulate Protein–Protein Interactions

- Protein concentration, which in turn are affected by expression levels and degradation rates;
- Protein affinity for proteins or other binding ligands;
- Ligands concentrations (substrates, ions, etc.);
- Presence of other proteins, nucleic acids, and ions;
- Electric fields around proteins.
- Occurrence of covalent modifications;

Structural Domains Involved in Protein–Protein Interactions

Proteins hold structural domains that allow their interaction with and bind to specific sequences on other proteins:

- Src homology 2 (SH2) domain: SH2 domains are structurally composed by three-stranded twisted beta sheet sandwiched flanked by two alpha-helices. The existence of a deep binding pocket with high affinity for phosphotyrosine, but not for phosphoserine or phosphothreonine, is essential for the recognition of tyrosine phosphorylated proteins, mainly autophosphorylated growth factor receptors. Growth factor receptor binding proteins and phospholipase Cγ are examples of proteins that have SH2 domains.
- Src homology 3 (SH3) domain: Structurally, SH3 domains are constituted by a beta barrel formed by two orthogonal beta sheets and three anti-parallel beta strands. These domains recognise proline enriched sequences, as polyproline type II helical structure (PXXP motifs) in cell signaling proteins like protein tyrosine kinases and the growth factor receptor bound protein 2 (Grb2).
- Phosphotyrosine-binding (PTB) domain: PTB domains interact with sequences that contain a phosphotyrosine group. These domains can be found in the insulin receptor substrate.
- LIM domain: LIM domains were initially identified in three homeodomain transcription factors (lin11, is11, and mec3). In addition to this homeodomain proteins and other proteins involved in development, LIM domains have also been identified in non-homeodomain proteins with relevant roles in cellular differentiation, association with cytoskeleton and senescence. These domains contain a tandem cysteine-rich Zn^{2+}-finger motif and embrace the consensus sequence CX2CX16-23HX2CX2CX2CX16-21CX2C/H/D. LIM domains bind to PDZ domains, bHLH transcription factors, and other LIM domains.
- Sterile alpha motif (SAM) domain: SAM domains are composed by five helices forming a compact package with a conserved hydrophobic core. These domains, which can be found in the Eph receptor and the stromal interaction molecule (STIM) for example, bind to non-SAM domain-containing proteins and they also appear to have the ability to bind RNA.
- PDZ domain: PDZ domains were first identified in three guanylate kinases: PSD-95, DlgA and ZO-1. These domains recognise carboxy-terminal tri-peptide motifs (S/TXV), other PDZ domains or LIM domains and bind them through a short peptide sequence that has a C-terminal hydrophobic residue. Some of the proteins identified as having PDZ domains are

scaffolding proteins or seem to be involved in ion receptor assembling and receptor-enzyme complexes formation.

- FERM domain

FERM domains contain basic residues capable of binding PtdIns (4,5)P_2. Talin and focal adhesion kinase (FAK) are two of the proteins that present FERM domains.

- Calponin homology (CH) domain: CH domains are mainly present in cytoskeletal proteins as parvin.
- Pleckstrin homology domain : Pleckstrin homology domains bind to phosphoinositides and acid domains in signaling proteins.
- WW domain: WW domains bind to proline enriched sequences.
- WSxWS motif: Found in cytokine receptors

Methods to Investigate Protein–Protein Interactions

There are a multitude of methods to detect them. Each of the approaches has its own strengths and weaknesses, especially with regard to the sensitivity and specificity of the method. The most conventional and widely used high-throughput methods are yeast two-hybrid screening and affinity purification coupled to mass spectrometry.

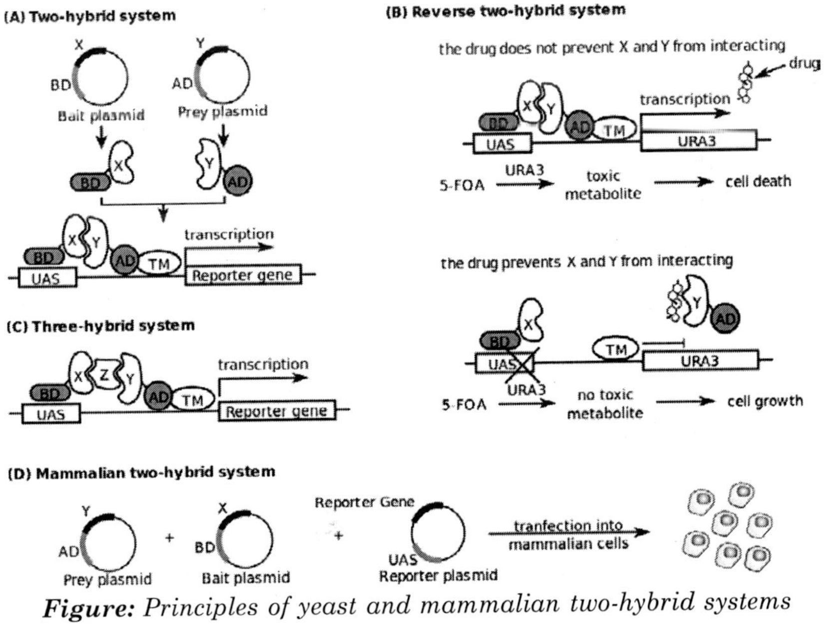

Figure: *Principles of yeast and mammalian two-hybrid systems*

Yeast Two-Hybrid Screening

This system was firstly described in 1989 by Fields and Song using *Saccharomyces cerevisiae* as biological model. Yeast two hybrid allows the identification of pairwise PPIs (binary method) *in vivo*, indicating non-specific tendencies towards sticky interactions.

Yeast cells are transfected with two plasmids: the bait (protein of interest fused with the DNA-binding domain of a yeast transcription factor, like Gal4), and the prey (a library of cDNA fragments linked to the activation domain of the transcription factor. Transcription of reporter genes does not occur unless bait and prey interact with each other and form a functional transcription factor. Thus, the interaction between proteins can be inferred by the presence of the products resultant of the reporter gene expression.

Despite its usefulness, the yeast two-hybrid system has limitations: specificity is relatively low; uses yeast as main host system, which can be a problem when studying other biological models; the number of PPIs identified is usually low because some transient PPIs are lost during purification steps; and, understates membrane proteins, for example. Limitations have been overcoming by the emergence of yeast two-hybrid variants, such as the membrane yeast two-hybrid (MYTH) and the split-ubiquitin system, which are not limited to interactions that occur in the nucleus; and, the bacterial two-hybrid system, performed in bacteria;

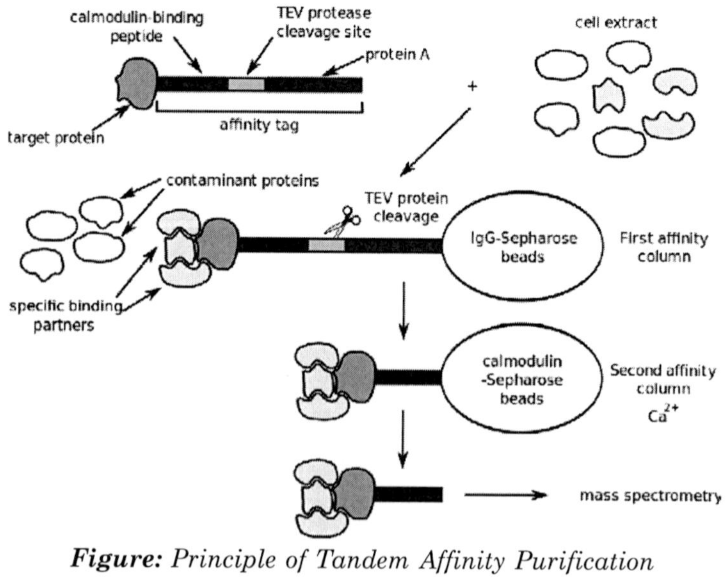

Figure: *Principle of Tandem Affinity Purification*

Affinity Purification Coupled to Mass Spectrometry

Affinity purification coupled to mass spectrometry mostly detects stable interactions and thus better indicates functional in vivo PPIs. This method starts by purification of the tagged protein, which is expressed in the cell usually at *in vivo* concentrations, and its interacting proteins (affinity purification). One of the most advantageous and widely used method to purify proteins with very low contaminating background is the tandem affinity purification, developed by Bertrand Seraphin and Mathias Mann and respective colleagues. PPIs can then be quantitatively and qualitatively analysed by mass spectrometry using different methods: chemical incorporation, biological or metabolic incorporation (SILAC), and label-free methods.

Other Potential Methods

Diverse techniques to identify PPIs have been emerging along with technology progression. These include co-immunoprecipitation, protein microarrays, analytical ultracentrifugation, light scattering, fluorescence spectroscopy, luminescence-based mammalian interactome mapping (LUMIER), resonance-energy transfer systems, mammalian protein–protein interaction trap, electro-switchable biosurfaces, surface plasmon resonance, protein-fragment complementation assay, and calorimetry.

Text Mining Methods

Recently text-mining methods were implemented to extract automatically protein–protein interactions from the literature. These methods generally detect binary relations between interacting proteins from individual sentences using machine learning and rule/pattern based information extraction and machine learning approaches.

Protein–Protein Interaction Databases

Large scale identification of PPIs generated hundreds of thousands interactions, which were collected together in specialised biological databases that are continuously updated in order to provide complete interactomes. The first of these databases was the Database of Interacting Proteins (DIP). Since that time, the number of public databases has been increasing. Databases can be subdivided into primary databases, meta-databases, and prediction databases.

- Primary databases collect information about published PPIs proven to exist via small-scale or large-scale experimental methods. Examples: DIP, Biomolecular Interaction Network Database (BIND), Biological General Repository for Interaction Datasets (BioGRID), Human Protein Reference Database (HPRD), IntAct Molecular Interaction Database, Molecular Interactions Database (MINT), MIPS Protein Interaction Resource on Yeast (MIPS-MPact), and MIPS Mammalian Protein–Protein Interaction Database (MIPS-MPPI).

- Meta-databases normally result from the integration of primary databases information, but can also collect some original data. Examples: Agile Protein Interaction DataAnalyzer (APID), The Microbial Protein Interaction Database (MPIDB), and Protein Interaction Network Analysis (PINA) platform.
- Prediction databases include many PPIs that are predicted using several techniques (main article). Examples: Michigan Molecular Interactions (MiMI), Human Protein–Protein Interaction Prediction Database (PIPs), Online Predicted Human Interaction Database (OPHID), Known and Predicted Protein–Protein Interactions (STRING), and Unified Human Interactome (UniHI).

Protein–Protein Interaction Networks

Information found in PPIs databases supports the construction of interaction networks. Although the PPI network of a given query protein can be represented in textbooks, diagrams of whole cell PPIs are frankly complex and difficult to generate.

One example of a manually produced molecular interaction map is the Kurt Kohn's 1999 map of cell cycle control. Drawing on Kohn's map, Schwikowski et al. in 2000 published a paper on PPIs in yeast, linking together 1,548 interacting proteins determined by two-hybrid screening. They used a layered graph drawing method to find an initial placement of the nodes and then improved the layout using a force-based algorithm. Bioinformatic tools have been developed to simplify the difficult task of visualize molecular interaction networks and complement them with other types of data. For instance, Cytoscape is an open-source software widely used and lots of plugins are currently available. Pajek software is advantageous for the visualization and analysis of very large networks.

The awareness of the major roles of PPIs in numerous physiological and pathological processes has been driving the challenge of unravel many interactomes. Examples of published interactomes are the thyroid specific DREAM interactome and the PP1α interactome in human brain.

Protein–Protein Interaction as Therapeutic Targets

Modulation of PPI is challenging and is receiving increasing attention by the scientific community. Several properties of PPI such as allosteric sites and hotspots, have been incorporated into drug-design strategies. The relevance of PPI as putative therapeutic targets for the development of new treatments is particularly evident in cancer, with several ongoing clinical trials within this area. The consensus among these promising targets is, nonetheless, denoted in the already available drugs on the market to treat a multitude of diseases. Examples are Titrobifan, inhibitor of the glycoprotein IIb/IIIa, used as a cardiovascular drug, and Maraviroc, inhibitor of the CCR5-gp120 interaction, used as anti-HIV drug. Recently, Amit Jaiswal and others were able to develop 30 peptides using protein–protein interaction studies to inhibit telomerase recruitment towards telomeres.

Protein–Protein Interaction Prediction

Protein–protein interaction prediction is a field combining bioinformatics and structural biology in an attempt to identify and catalog physical interactions between pairs or groups of proteins. Understanding protein–protein interactions is important for the investigation of intracellular signaling pathways, modelling of protein complex structures and for gaining insights into various biochemical processes. Experimentally, physical interactions between pairs of proteins can be inferred from a variety of experimental techniques, including yeast two-hybrid systems, protein-fragment complementation assays (PCA), affinity purification/mass spectrometry, protein microarrays, fluorescence resonance energy transfer (FRET), and Microscale Thermophoresis (MST). Efforts to experimentally determine the interactome of numerous species are ongoing, and a number of computational methods for interaction prediction have been developed in recent years.

Methods

Proteins that interact are more likely to co-evolve, therefore, it is possible to make inferences about interactions between pairs of proteins based on their phylogenetic distances. It has also been observed in some cases that pairs of interacting proteins have fused orthologues in other organisms. In addition, a number of bound protein complexes have been structurally solved and can be used to identify the residues that mediate the interaction so that similar motifs can be located in other organisms.

Phylogenetic Profiling

Phylogenetic profiling finds pairs of protein families with similar patterns of presence or absence across large numbers of species. This method is based on the hypothesis that potentially interacting proteins should co-evolve and should have orthologs in closely related species. That is, proteins that form complexes or are part of a pathway should be present simultaneously in order for them to function. A phylogenetic profile is constructed for each protein under investigation. The profile is basically a record of whether the protein is present in certain genomes. If two proteins are found to be present and absent in the same genomes, those proteins are deemed likely to be functionally related. A similar method can be applied to protein domains, where profiles are constructed for domains to determine if there are domain interactions. Some drawbacks with the phylogenetic profile methods are that they are computationally expensive to perform, they rely on homology detection between distant organisms, and they only identify if the proteins being investigated are functionally related (part of complex or in same pathway) and not if they have direct interactions.

Prediction of Co-Evolved Protein Pairs based on Similar Phylogenetic Trees

It was observed that the phylogenetic trees of ligands and receptors were often more similar than due to random chance. This is likely because they faced similar selection pressures and co-evolved. This method uses the phylogenetic trees of protein pairs to determine if interactions exist. To do this, homologs of the proteins of interest are found (using a sequence search tool such as BLAST) and multiple-sequence alignments are done (with alignment tools such as Clustal) to build distance matrices for each of the proteins of interest. The distance matrices should then be used to build phylogenetic trees. However, comparisons between phylogenetic trees are difficult, and current methods circumvent this by simply comparing distance matrices. The distance matrices of the proteins are used to calculate a correlation coefficient, in which a larger value corresponds to co-evolution. The benefit of comparing distance matrices instead of phylogenetic trees is that the results do not depend on the method of tree building that was used.

The downside is that difference matrices are not perfect representations of phylogenetic trees, and inaccuracies may result from using such a shortcut. Another factor worthy of note is that there are background similarities between the phylogenetic trees of any protein, even ones that do not interact. If left unaccounted for, this could lead to a high false-positive rate. For this reason, certain methods construct a background tree using 16S rRNA sequences which they use as the canonical tree of life. The distance matrix constructed from this tree of life is then subtracted from the distance matrices of the proteins of interest. However, because RNA distance matrices and DNA distance matrices have different scale, presumably because RNA and DNA have different mutation rates, the RNA matrix needs to be rescaled before it can be subtracted from the DNA matrices. By using molecular clock proteins, the scaling coefficient for protein distance/RNA distance can be calculated. This coefficient is used to rescale the RNA matrix.

Rosetta Stone Method

A Rosetta stone protein is a protein chain composed of two fused proteins. It is observed that proteins or domains that interact with one another tend to have homologs in other genomes that are fused into a Rosetta stone protein. It is thought that the fusion helps optimize the co-expression of these proteins. The most obvious drawback of this method is that there are many protein interactions that cannot be discovered this way; it relies on the presence of Rosetta stone proteins. Also, like phylogenetic profile methods, the Rosetta stone method does not necessarily find interacting proteins, as there can be other reasons for the fusion of two proteins.

Classification Methods

Classification methods use data to train a program (classifier) to distinguish positive examples of interacting protein/domain pairs with negative examples of non-interacting pairs. Popular classifiers used are Random Forest Decision (RFD) and Support Vector Machines. RFD produces results based on the domain composition of interacting and non-interacting protein pairs. When given a protein pair to classify, RFD first creates a representation of the protein pair in a vector. The vector contains all the domain types used to train RFD, and for each domain type the vector also contains a value of 0, 1, or 2. If the protein pair does not contain a certain domain, then the value for that domain is 0. If one of the proteins of the pair contains the domain, then the value is 1. If both proteins contain the domain, then the value is 2. Using training data, RFD constructs a decision forest, consisting of many decision trees. Each decision tree evaluates several domains, and based on the presence or absence of interactions in these domains, makes a decision as to if the protein pair interacts. The vector representation of the protein pair is evaluated by each tree to determine if they are an interacting pair or a non-interacting pair. The forest tallies up all the input from the trees to come up with a final decision. The strength of this method is that it does not assume that domains interact independent of each other. This makes it so that multiple domains in proteins can be used in the prediction. This is a big step up from previous methods which could only predict based on a single domain pair. The limitation of this method is that it relies on the training dataset to produce results. Thus, usage of different training datasets could influence the results.

Inference of Interactions from Homologous Structures

This group of methods makes use of known protein complex structures to predict and structurally model interactions between query protein sequences. The prediction process generally starts by employing a sequence based method (e.g. Interolog) to search for protein complex structures that are homologous to the query sequences. These known complex structures are then used as templates to structurally model the interaction between query sequences. This method has the advantage of not only inferring protein interactions but also suggests models of how proteins interact structurally, which can provide some insights into the atomic level mechanism of that interaction. On the other hand, the ability for these methods to make a prediction is constrained by a limited number of known protein complex structures.

Association Methods

Association methods look for characteristic sequences or motifs that can help distinguish between interacting and non-interacting pairs. A classifier is trained by looking for sequence-signature pairs where one protein contains one sequence-signature, and its interacting partner contains another sequence-signature. They look specifically for sequence-signatures that are found together more often than by chance. This uses a log-odds score which is computed as log2(Pij/PiPj), where Pij is the observed frequency of domains i and j occurring in one protein pair; Pi and Pj are the background frequencies of domains i and j in the data. Predicted domain interactions are those with positive log-odds scores and also having several occurrences within the database. The downside with this method is that it looks at each pair of interacting domains separately, and it assumes that they interact independently of each other.

Identification of Structural Patterns

This method builds a library of known protein–protein interfaces from the PDB, where the interfaces are defined as pairs of polypeptide fragments that are below a threshold slightly larger than the Van der Waals radius of the atoms involved. The sequences in the library are then clustered based on structural alignment and redundant sequences are eliminated. The residues that have a high (generally >50%) level of frequency for a given position are considered hotspots. This library is then used to identify potential interactions between pairs of targets, providing that they have a known structure (i.e. present in the PDB).

Bayesian Network Modelling

Bayesian methods integrate data from a wide variety of sources, including both experimental results and prior computational predictions, and use these features to assess the likelihood that a particular potential protein interaction is a true positive result. These methods are useful because experimental procedures, particularly the yeast two-hybrid experiments, are extremely noisy and produce many false positives, while the previously mentioned computational methods can only provide circumstantial evidence that a particular pair of proteins might interact.

Domain-Pair Exclusion Analysis

The domain-pair exclusion analysis detects specific domain interactions that are hard to detect using Bayesian methods. Bayesian methods are good at detecting nonspecific promiscuous interactions and not very good at detecting rare specific interactions. The domain-pair exclusion analysis method calculates an E-score which measures if two domains interact. It is calculated as log (probability that the two proteins interact given that the domains interact/probability that the two proteins interact given that the domains don't interact). The probabilities required in the formula are calculated using an Expectation Maximization procedure, which is a method for estimating parameters in statistical models. High E-scores indicate that the two domains are likely to interact, while low scores indicate that other domains form the protein pair are more likely to be responsible for the interaction. The drawback with this method is that it does not take into account false positives and false negatives in the experimental data.

Supervised Learning Problem

The problem of PPI prediction can be framed as a supervised learning problem. In this paradigm the known protein interactions supervise the estimation of a function that can predict whether an interaction exists or not between two proteins given data about the proteins (e.g., expression levels of each gene in different experimental conditions, location information, phylogenetic profile, etc.).

Relationship to Docking Methods

The field of protein–protein interaction prediction is closely related to the field of protein–protein docking, which attempts to use geometric and steric considerations to fit two proteins of known structure into a bound complex. This is a useful mode of inquiry in cases where both proteins in the pair have known structures and are known (or at least strongly suspected) to interact, but since so many proteins do not have experimentally determined structures, sequence-based interaction prediction methods are especially useful in conjunction with experimental studies of an organism's interactome.

Chapter 3

Sequence Analysis

In bioinformatics, sequence analysis is the process of subjecting a DNA, RNA or peptide sequence to any of a wide range of analytical methods to understand its features, function, structure, or evolution. Methodologies used include sequence alignment, searches against biological databases, and others. Since the development of methods of high-throughput production of gene and protein sequences, the rate of addition of new sequences to the databases increased exponentially. Such a collection of sequences does not, by itself, increase the scientist's understanding of the biology of organisms. However, comparing these new sequences to those with known functions is a key way of understanding the biology of an organism from which the new sequence comes. Thus, sequence analysis can be used to assign function to genes and proteins by the study of the similarities between the compared sequences. Nowadays, there are many tools and techniques that provide the sequence comparisons (sequence alignment) and analyze the alignment product to understand its biology.

Sequence analysis in molecular biology includes a very wide range of relevant topics:

1. The comparison of sequences in order to find similarity, often to infer if they are related (homologous)
2. Identification of intrinsic features of the sequence such as active sites, post translational modification sites, gene-structures, reading frames, distributions of introns and exons and regulatory elements
3. Identification of sequence differences and variations such as point mutations and single nucleotide polymorphism (SNP) in order to get the genetic marker.

4. Revealing the evolution and genetic diversity of sequences and organisms
5. Identification of molecular structure from sequence alone

In chemistry, sequence analysis comprises techniques used to determine the sequence of a polymer formed of several monomers. In molecular biology and genetics, the same process is called simply "sequencing".In marketing, sequence analysis is often used in analytical customer relationship management applications, such as NPTB models (Next Product to Buy). In sociology, sequence methods are increasingly used to study life-course and career trajectories, patterns of organisational and national development, conversation and interaction structure, and the problem of work/family synchrony. This body of research has given rise to the emerging subfield of social sequence analysis.

History

Since the very first sequences of the insulin protein was characterised by Fred Sanger in 1951 biologists have been trying to use this knowledge to understand the function of molecules. According to Michael Levitt, sequence analysis was born in the period from 1969-1977. In 1969 the analysis of sequences of transfer RNAs were used to infer residue interactions from correlated changes in the nucleotide sequences giving rise to a model of the tRNA secondary structure. In 1970, Saul B. Needleman and Christian D. Wunsch published the first computer algorithm for aligning two sequences. Over this time developments in obtaining nucleotide sequence greatly improved leading to the publication of the first complete genome of a bacteriophage in 1977.

Sequence Alignment

There are millions of protein and nucleotide sequences known. These sequences fall into many groups of related sequences known as protein families or gene families. Relationships between these sequences are usually discovered by aligning them together and assigning this alignment a score. There are two main types of sequence alignment. Pair-wise sequence alignment only compares two sequences at a time and multiple sequence alignment compares many sequences in one go. Two important algorithms for aligning pairs of sequences are the Needleman-Wunsch algorithm and the Smith-Waterman algorithm. Popular tools for sequence alignment include:

- Pair-wise alignment - BLAST
- Multiple alignment - ClustalW, PROBCONS, MUSCLE, MAFFT, and T-Coffee.

A common use for pairwise sequence alignment is to take a sequence of interest and compare it to all known sequences in a database to identify homologous sequences. In general the matches in the database are ordered to show the most closely related sequences first followed by sequences with diminishing similarity. These matches are usually reported with a measure of statistical significance such as an Expectation value.

Profile Comparison

In 1987, Michael Gribskov, Andrew McLachlan, and David Eisenberg introduced the method of profile comparison for identifying distant similarities between proteins. Rather than using a single sequence, profile methods use a multiple sequence alignment to encode a profile which contains information about the conservation level of each residue. These profiles can then be used to search collections of sequences to find sequences that are related. Profiles are also known as Position Specific Scoring Matrices (PSSMs). In 1993, a probabilistic interpretation of profiles was introduced by David Haussler and colleagues using hidden Markov models. These models have become known as profile-HMMs.

In recent years, methods have been developed that allow the comparison of profiles directly to each other. These are known as profile-profile comparison methods.

Methodology

The tasks that lie in the space of sequence analysis are often non-trivial to resolve and require the use of relatively complex approaches. Of the many types of methods used in practice, the most popular include:

- Dynamic programing
- Artificial Neural Network
- Hidden Markov Model
- Support Vector Machine
- Clustering
- Bayesian Network
- Regression Analysis
- Sequence mining
- Alignment-free sequence analysis

Sequence Assembly

In bioinformatics, sequence assembly refers to aligning and merging fragments of a much longer DNA sequence in order to reconstruct the original sequence. This is needed as DNA sequencing technology cannot read whole genomes in one go, but rather reads small pieces of between 20 and 30000 bases, depending on the technology used. Typically the short fragments, called reads, result from shotgun sequencing genomic DNA, or gene transcript (ESTs).

The problem of sequence assembly can be compared to taking many copies of a book, passing each of them through a shredder with a different cutter, and piecing the text of the book back together just by looking at the shredded pieces. Besides the obvious difficulty of this task, there are some extra practical issues: the original may have many repeated paragraphs, and some shreds may be modified during shredding to have typos. Excerpts from another book may also be added in, and some shreds may be completely unrecognisable.

Genome Assemblers

The first sequence assemblers began to appear in the late 1980s and early 1990s as variants of simpler sequence alignment programs to piece together vast quantities of fragments generated by automated sequencing instruments called DNA sequencers. As the sequenced organisms grew in size and complexity (from small viruses over plasmids to bacteria and finally eukaryotes), the assembly programs used in these genome projects needed increasingly sophisticated strategies to handle:

- terabytes of sequencing data which need processing on computing clusters;
- identical and nearly identical sequences (known as *repeats*) which can, in the worst case, increase the time and space complexity of algorithms exponentially;
- errors in the fragments from the sequencing instruments, which can confound assembly.

Faced with the challenge of assembling the first larger eukaryotic genomes—the fruit fly Drosophila melanogaster in 2000 and the human genome just a year later,—scientists developed assemblers like Celera Assembler and Arachne able to handle genomes of 100-300 million base pairs.

Subsequent to these efforts, several other groups, mostly at the major genome sequencing centres, built large-scale assemblers, and an open source effort known as AMOS was launched to bring together all the innovations in genome assembly technology under the open source framework.

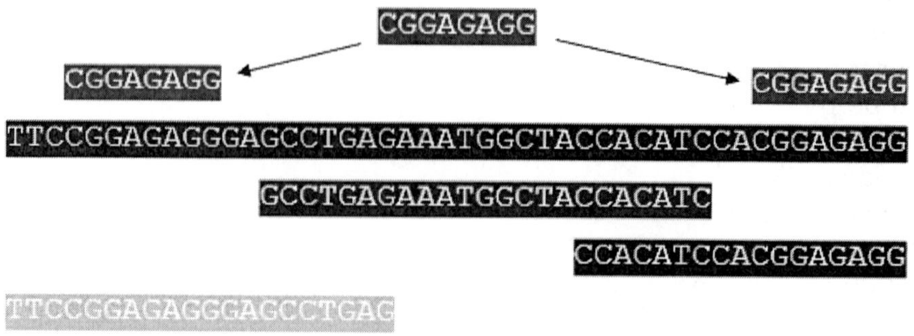

Figure: *Sample sequence showing how a sequence assembler would take fragments and match by overlaps. Image also shows the potential problem of repeats in the sequence.*

EST Assemblers

Expressed Sequence Tag or EST assembly differs from genome assembly in several ways. The sequences for EST assembly are the transcribed mRNA of a cell and represent only a subset of the whole genome. At a first glance, underlying algorithmical problems differ between genome and EST assembly. For instance, genomes often have large amounts of repetitive sequences, mainly in the inter-genic parts. Since ESTs represent gene transcripts, they will not contain these repeats. On the other hand, cells tend to have a certain number of genes that are constantly expressed in very high numbers (housekeeping genes), which again leads to the problem of similar sequences present in high numbers in the data set to be assembled.

Furthermore, genes sometimes overlap in the genome (sense-antisense transcription), and should ideally still be assembled separately. EST assembly is also complicated by features like (cis-) alternative splicing, trans-splicing, single-nucleotide polymorphism, recoding, and post-transcriptional modification.

De-Novo vs. Mapping Assembly

In sequence assembly, two different types can be distinguished:

1. de-novo: assembling short reads to create full-length (sometimes novel) sequences
2. mapping: assembling reads against an existing backbone sequence, building a sequence that is similar but not necessarily identical to the backbone sequence

In terms of complexity and time requirements, de-novo assemblies are orders of magnitude slower and more memory intensive than mapping assemblies. This is mostly due to the fact that the assembly algorithm needs to compare every read with every other read (an operation that has a naive time complexity of $O(n^2)$; using a hash this can be reduced significantly). Referring to the comparison drawn to shredded books in the introduction: while for mapping assemblies one would have a very similar book as template (perhaps with the names of the main characters and a few locations changed), the de-novo assemblies are more hardcore in a sense as one would not know beforehand whether this would become a science book, a novel, a catalogue, or even several books. Also, every shred would be compared with every other shred.

Influence of Technological Changes

The complexity of sequence assembly is driven by two major factors: the number of fragments and their lengths. While more and longer fragments allow better identification of sequence overlaps, they also pose problems as the underlying algorithms show quadratic or even exponential complexity behaviour to both number of fragments and their length. And while shorter sequences are faster to align, they also complicate the layout phase of an assembly as shorter reads are more difficult to use with repeats or near identical repeats.

In the earliest days of DNA sequencing, scientists could only gain a few sequences of short length (some dozen bases) after weeks of work in laboratories. Hence, these sequences could be aligned in a few minutes by hand.

In 1975, the Dideoxy termination method (also known as *Sanger sequencing*) was invented and until shortly after 2000, the technology was improved up to a point where fully automated machines could churn out sequences in a highly parallelised mode 24 hours a day. Large genome centres around the world housed complete farms of these sequencing machines, which in turn led to the necessity of assemblers to be optimised for sequences from whole-genome shotgun sequencing projects where the reads

- are about 800–900 bases long
- contain sequencing artifacts like sequencing and cloning vectors
- have error rates between 0.5 and 10%

With the Sanger technology, bacterial projects with 20,000 to 200,000 reads could easily be assembled on one computer. Larger projects, like the human genome with approximately 35 million reads, needed large computing farms and distributed computing.

By 2004 / 2005, pyrosequencing had been brought to commercial viability by 454 Life Sciences. This new sequencing method generated reads much shorter than those of Sanger sequencing: initially about 100 bases, now 400-500 bases. Its much higher throughput and lower cost (compared to Sanger sequencing) pushed the adoption of this technology by genome centres, which in turn pushed development of sequence assemblers that could efficiently handle the read sets.

The sheer amount of data coupled with technology-specific error patterns in the reads delayed development of assemblers; at the beginning in 2004 only the Newbler assembler from 454 was available. Released in mid-2007, the hybrid version of the MIRA assembler by Chevreux et al. was the first freely available assembler that could assemble 454 reads as well as mixtures of 454 reads and Sanger reads. Assembling sequences from different sequencing technologies was subsequently coined *hybrid assembly*.

From 2006, the Illumina (previously Solexa) technology has been available and can generate about 100 million reads per run on a single sequencing machine. Compare this to the 35 million reads of the human genome project which needed several years to be produced on hundreds of sequencing machines. Illumina was initially limited to a length of only 36 bases, making it less suitable for de novo assembly (such as de novo transcriptome assembly), but newer iterations of the technology achieve read lengths above 100 bases from both ends of a 3-400bp clone. Announced at the end of 2007, the SHARCGS assembler by Dohm et al. was the first published assembler that was used for an assembly with Solexa reads. It was quickly followed by a number of others.

Later, new technologies like SOLiD from Applied Biosystems, Ion Torrent and SMRT were released and new technologies (e.g. Nanopore sequencing) continue to emerge.

Greedy Algorithm

Given a set of sequence fragments the object is to find the shortest common supersequence.

1. Ñalculate pairwise alignments of all fragments.
2. Choose two fragments with the largest overlap.
3. Merge chosen fragments.
4. Repeat step 2 and 3 until only one fragment is left.

The result is a suboptimal solution to the problem.

Available Assemblers

The following table lists assemblers that have a de-novo assembly capability on at least one of the supported technologies.

Name	*Type*	*Technologies*	*Author*	*Presented / Last updated*	*Licence**	*Homepage*
ABySS	(large) genomes	Solexa, SOLiD	Simpson, J. et al.	2008 / 2011	NC-A	link
ALLPATHS-LG	(large) genomes	Solexa, SOLiD	Gnerre, S. et al.	2011	OS	link
AMOS	genomes	Sanger, 454	Salzberg, S. et al.	2002? / 2008?	OS	link
Arapan-M	Medium Genomes (e.g. E.coli)	All	Sahli, M. & Shibuya, T.	2011 / 2012	OS	link
Arapan-S	Small Genomes (Viruses and Bacteria)	All	Sahli, M. & Shibuya, T.	2011 / 2012	OS	link
Celera WGA Assembler / CABOG	(large) genomes	Sanger, 454, Solexa	Myers, G. et al.; Miller G. et al.	2004 / 2010	OS	link
CLC Genomics Workbench & CLC Assembly Cell	genomes	Sanger, 454, Solexa, SOLiD	CLC bio	2008 / 2010 / 2011	C	link
Cortex	genomes	Solexa, SOLiD	Iqbal, Z. *et al.*	2011	OS	link
DNA Baser	genomes	Sanger, 454	Heracle BioSoft SRL	01.2014	C	[www.DnaBaser.com link]
DNA Dragon	genomes	Illumina, SOLiD, Complete Genomics, 454, Sanger	SequentiX	2011	C	link
DNAnexus	genomes	Illumina, SOLiD, Complete Genomics	DNAnexus	2011	C	link
Edena	genomes	Illumina	D. Hernandez, P. François, L. Farinelli, M. Osteras, and J. Schrenzel.	2008/2013	OS	link
Euler	genomes	Sanger, 454 (,Solexa ?)	Pevzner, P. et al.	2001 / 2006?	(C / NC-A?)	link
Euler-sr	genomes	454, Solexa	Chaisson, MJ. et al.	2008	NC-A	link
Fermi	(large) genomes	Illumina	Li, H.	2012	OS	link
Forge	(large) genomes, EST, metagenomes	454, Solexa, SOLID, Sanger	Platt, DM, Evers, D.	2010	OS	link
Geneious	genomes	Sanger, 454, Solexa, Ion Torrent, Complete Genomics, PacBio, Oxford Nanopore, Illumina	Biomatters Ltd	2009 / 2013	C	link
Graph Constructor	(large) genomes	Sanger, 454, Solexa, SOLiD	Convey Computer Corporation	2011	C	link
IDBA (Iterative De Bruijn graph short read Assembler)	(large) genomes	Sanger,454,Solexa	Yu Peng, Henry C. M. Leung, Siu-Ming Yiu, Francis Y. L. Chin	2010	(C / NC-A?)	link
LIGR Assembler (derived from TIGR Assembler)	genomic	Sanger	-	2009/ 2012	OS	link
MaSuRCA (Maryland Super Read - Celera Assembler)	(large) genomes	Sanger, Illumina, 454	Aleksey Zimin, Guillaume Marçais, Daniela Puiu, Michael Roberts, Steven L. Salzberg, James A. Yorke	2012 / 2013	OS	link
MIRA (Mimicking Intelligent Read Assembly)	genomes, ESTs	Sanger, 454, Solexa	Chevreux, B.	1998 / 2011	OS	link
NextGENe	(small genomes?)	454, Solexa, SOLiD	Softgenetics	2008	C	link
Newbler	genomes, ESTs	454, Sanger	454/Roche	2009	C	link
PADENA	genomes	454, Sanger	454/Roche	2010	OS	link
PASHA	(large) genomes	Illumina	Liu, Schmidt, Maskell	2011	OS	link
Phrap	genomes	Sanger, 454, Solexa	Green, P	1994 / 2008	C / NC-A	link
TIGR Assembler	genomic	Sanger	-	1995 / 2003	OS	link
Ray	genomes	Illumina, mix of Illumina and 454, paired or not	Sébastien Boisvert, François Laviolette & Jacques Corbeil.	2010	OS [GNU General Public License]	link
Sequencher	genomes	traditional and next generation sequence data	Gene Codes Corporation	1991 / 2009 / 2011	C	link
SeqMan NGen	(large) genomes, exomes, transcriptomes, metagenomes, ESTs	Illumina, ABI SOLiD, Roche 454, Ion Torrent, Solexa, Sanger	DNASTAR	2007 / 2011	C	link
SGA	(large) genomes	Illumina, Sanger (Roche 454?, Ion Torrent?)	Simpson, J.T. et al.	2011 / 2012	OS	link
SHARCGS	(small) genomes	Solexa	Dohm et al.	2007 / 2007	OS	link
SOPRA	genomes	Illumina, SOLiD, Sanger, 454	Dayarian, A. et al.	2010 / 2011	OS	link
SparseAssembler	(large) genomes	Illumina, 454, Ion torrent	Ye, C. et al.	2012 / 2012	OS	link
SSAKE	(small) genomes	Solexa (SOLiD? Helicos?)	Warren, R. et al.	2007 / 2007	OS	link
SOAPdenovo	genomes	Solexa	Li, R. et al.	2009 / 2009	OS	link
SPAdes	(small) genomes, single-cell	Illumina, Solexa	Bankevich, A et al.	2012 / 2013	OS	link
Staden gap4 package	BACs (, small genomes?)	Sanger	Staden et al.	1991 / 2008	OS	link
Taipan	(small) genomes	Illumina	Schmidt, B. *et al.*	2009	OS	link
VCAKE	(small) genomes	Solexa (SOLiD?, Helicos?)	Jeck, W. et al.	2007 / 2007	OS	link
Phusion assembler	(large) genomes	Sanger	Mullikin JC, *et al.*	2003	OS	link
Quality Value Guided SRA (QSRA)	genomes	Sanger, Solexa	Bryant DW, *et al.*	2009	OS	link
Velvet	(small) genomes	Sanger, 454, Solexa, SOLiD	Zerbino, D. et al.	2007 / 2009	OS	link

*Licences: OS = Open Source; C = Commercial; C / NC-A = Commercial but free for non-commercial and academics; Brackets = unclear, but most likely C / NC-A

Threading (Protein Sequence)

Protein threading, also known as fold recognition, is a method of protein modelling (i.e. computational protein structure prediction) which is used to model those proteins which have the same fold as proteins of known structures, but do not have homologous proteins with known structure. It differs from the homology modelling method of structure prediction as it (protein threading) is used for proteins which do not have their homologous protein structures deposited in the Protein Data Bank (PDB), whereas homology modelling is used for those proteins which do. Threading works by using statistical knowledge of the relationship between the structures deposited in the PDB and the sequence of the protein which one wishes to model.

The prediction is made by "threading" (i.e. placing, aligning) each amino acid in the target sequence to a position in the template structure, and evaluating how well the target fits the template. After the best-fit template is selected, the structural model of the sequence is built based on the alignment with the chosen template. Protein threading is based on two basic observations: that the number of different folds in nature is fairly small (approximately 1300); and that 90% of the new structures submitted to the PDB in the past three years have similar structural folds to ones already in the PDB.

Classification of Protein Structure

The Structural Classification of Proteins (SCOP) database provides a detailed and comprehensive description of the structural and evolutionary relationships of known structure. Proteins are classified to reflect both structural and evolutionary relatedness. Many levels exist in the hierarchy, but the principal levels are family, superfamily and fold, as described below.

Family (Clear Evolutionary Relationship): Proteins clustered together into families are clearly evolutionarily related. Generally, this means that pairwise residue identities between the proteins are 30% and greater. However, in some cases similar functions and structures provide definitive evidence of common descent in the absence of high sequence identity; for example, many globins form a family though some members have sequence identities of only 15%.

Superfamily (Probable Common Evolutionary Origin): Proteins that have low sequence identities, but whose structural and functional features suggest that a common evolutionary origin is probable, are placed together in superfamilies. For example, actin, the ATPase domain of the heat shock protein, and hexakinase together form a superfamily.

Fold (Major Structural Similarity): Proteins are defined as having a common fold if they have the same major secondary structures in the same arrangement and with the same topological connections. Different proteins with the same fold often have peripheral elements of secondary structure and turn regions that differ in size and conformation. In some cases, these differing peripheral regions may comprise half the structure. Proteins placed together in the same fold category may not have a common evolutionary origin: the structural similarities could arise just from the physics and chemistry of proteins favouring certain packing arrangements and chain topologies.

Method

A general paradigm of protein threading consists of the following four steps:

The Construction of a Structure Template Database: Select protein structures from the protein structure databases as structural templates. This generally involves selecting protein structures from databases such as PDB, FSSP, SCOP, or CATH, after removing protein structures with high sequence similarities.

The Design of the Scoring Function: Design a good scoring function to measure the fitness between target sequences and templates based on the knowledge of the known relationships between the structures and the sequences. A good scoring function should contain mutation potential, environment fitness potential, pairwise potential, secondary structure compatibilities, and gap penalties. The quality of the energy function is closely related to the prediction accuracy, especially the alignment accuracy.

Threading Alignment: Align the target sequence with each of the structure templates by optimizing the designed scoring function. This step is one of the major tasks of all threading-based structure prediction programs that take into account the pairwise contact potential; otherwise, a dynamic programming algorithm can fulfill it.

Threading Prediction: Select the threading alignment that is statistically most probable as the threading prediction. Then construct a structure model for the target by placing the backbone atoms of the target sequence at their aligned backbone positions of the selected structural template.

Comparison with Homology Modelling

Homology modelling and protein threading are both template-based methods and there is no rigorous boundary between them in terms of prediction techniques. But the protein structures of their targets are different. Homology modelling is for those targets which have homologous proteins with known structure (usually/maybe of same family), while protein threading is for those targets with only fold-level homology found. In other words, homology modelling is for "easier" targets and protein threading is for "harder" targets.

Homology modelling treats the template in an alignment as a sequence, and only sequence homology is used for prediction. Protein threading treats the template in an alignment as a structure, and both sequence and structure information extracted from the alignment are used for prediction. When there is no significant homology found, protein threading can make a prediction based on the structure information. That also explains why protein threading may be more effective than homology modelling in many cases.

In practice, when the sequence identity in a sequence sequence alignment is low (i.e. <25%), homology modelling may not produce a significant prediction. In this case, if there is distant homology found for the target, protein threading can generate a good prediction.

More about Threading

Fold recognition methods can be broadly divided into two types: 1, those that derive a 1-D profile for each structure in the fold library and align the target sequence to these profiles; and 2, those that consider the full 3-D structure of the protein template. A simple example of a profile representation would be to take each amino acid in the structure and simply label it according to whether it is buried in the core of the protein or exposed on the surface. More elaborate profiles might take into account the local secondary structure (e.g. whether the amino acid is part of an alpha helix) or even evolutionary information (how conserved the amino acid is).

In the 3-D representation, the structure is modelled as a set of inter-atomic distances, i.e. the distances are calculated between some or all of the atom pairs in the structure. This is a much richer and far more flexible description of the structure, but is much harder to use in calculating an alignment. The profile-based fold recognition approach was first described by Bowie, Lüthy and Eisenberg in 1991. The term *threading* was first coined by Jones, Taylor and Thornton in 1992, and originally referred specifically to the use of a full 3-D structure atomic representation of the protein template in fold recognition. Today, the terms threading and fold recognition are frequently (though somewhat incorrectly) used interchangeably.

Fold recognition methods are widely used and effective because it is believed that there are a strictly limited number of different protein folds in nature, mostly as a result of evolution but also due to constraints imposed by the basic physics and chemistry of polypeptide chains. There is, therefore, a good chance (currently 70-80%) that a protein which has a similar fold to the target protein has already been studied by X-ray crystallography or nuclear magnetic resonance (NMR) spectroscopy and can be found in the PDB. Currently there are nearly 1300 different protein folds known, but new folds are still being discovered every year due in significant part to the ongoing structural genomics projects.

Many different algorithms have been proposed for finding the correct threading of a sequence onto a structure, though many make use of dynamic programming in some form. For full 3-D threading, the problem of identifying the best alignment is very difficult (it is an NP-hard problem for some models of threading). Researchers have made use of many combinatorial optimization methods such as Conditional random fields, simulated annealing, branch and bound and linear programming, searching to arrive at heuristic solutions.

It is interesting to compare threading methods to methods which attempt to align two protein structures (protein structural alignment), and indeed many of the same algorithms have been applied to both problems.

Protein Threading Software

- HHpred is a popular threading server which runs HHsearch, a widely used software for remote homology detection based on pairwise comparison of hidden Markov models.
- RAPTOR (software) is an integer programming based protein threading software. It has been replaced by a new protein threading program RaptorX / software for protein modelling and analysis, which employs probabilistic graphical models and statistical inference to both single template and multi-template based protein threading. RaptorX significantly outperforms RAPTOR and is especially good at aligning proteins with sparse sequence profile. The RaptorX server is free to public.
- Phyre is a popular threading server combining HHsearch with *ab initio* and multiple-template modelling.
- MUSTER is a standard threading algorithm based on dynamic programming and sequence profile-profile alignment. It also combines multiple structural resources to assist the sequence profile alignment.
- SPARKS X is a probabilistic-based sequence-to-structure matching between predicted one-dimensional structural properties of query and corresponding native properties of templates.

- BioShell is a threading algorithm using optimized profile-to-profile dynamic programming algorithm combined with predicted secondary structure.

Combinatorics

Combinatorics is a branch of mathematics concerning the study of finite or countable discrete structures. Aspects of combinatorics include counting the structures of a given kind and size (enumerative combinatorics), deciding when certain criteria can be met, and constructing and analyzing objects meeting the criteria (as in combinatorial designs and matroid theory), finding "largest", "smallest", or "optimal" objects (extremal combinatorics and combinatorial optimization), and studying combinatorial structures arising in an algebraic context, or applying algebraic techniques to combinatorial problems (algebraic combinatorics).

Combinatorial problems arise in many areas of pure mathematics, notably in algebra, probability theory, topology, and geometry, and combinatorics also has many applications in mathematical optimization, computer science, ergodic theory and statistical physics. Many combinatorial questions have historically been considered in isolation, giving an *ad hoc* solution to a problem arising in some mathematical context. In the later twentieth century, however, powerful and general theoretical methods were developed, making combinatorics into an independent branch of mathematics in its own right. One of the oldest and most accessible parts of combinatorics is graph theory, which also has numerous natural connections to other areas. Combinatorics is used frequently in computer science to obtain formulas and estimates in the analysis of algorithms.

A mathematician who studies combinatorics is called a combinatorialist or a combinatorist.

History

Basic combinatorial concepts and enumerative results appeared throughout the ancient world. In 6th century BCE, ancient Indian physician Sushruta asserts in Sushruta Samhita that 63 combinations can be made out of 6 different tastes, taken one at a time, two at a time, etc., thus computing all $2^6 - 1$ possibilities. Greek historian Plutarch discusses an argument between Chrysippus (3rd century BCE) and Hipparchus (2nd century BCE) of a rather delicate enumerative problem, which was later shown to be related to Schröder numbers. In the *Ostomachion*, Archimedes (3rd century BCE) considers a tiling puzzle.

In the Middle Ages, combinatorics continued to be studied, largely outside of the European civilization. The Indian mathematician Mahâvîra (c. 850) provided formulae for the number of permutations and combinations, and these formulas may have been familiar to Indian mathematicians as early as the 6th century CE. The philosopher and astronomer Rabbi Abraham ibn Ezra (c. 1140) established the symmetry of binomial coefficients, while a closed formula was obtained later by the talmudist and mathematician Levi ben Gerson (better known as Gersonides), in 1321. The arithmetical triangle— a graphical diagram showing relationships among the binomial coefficients— was presented by mathematicians in treatises dating as far back as the 10th century, and would eventually become known as Pascal's triangle. Later, in Medieval England, campanology provided examples of what is now known as Hamiltonian cycles in certain Cayley graphs on permutations.

During the Renaissance, together with the rest of mathematics and the sciences, combinatorics enjoyed a rebirth. Works of Pascal, Newton, Jacob Bernoulli and Euler became foundational in the emerging field. In modern times, the works of J. J. Sylvester (late 19th century) and Percy MacMahon (early 20th century) laid the foundation for enumerative and algebraic combinatorics. Graph theory also enjoyed an explosion of interest at the same time, especially in connection with the four colour problem.

In the second half of 20th century, combinatorics enjoyed a rapid growth, which led to establishment of dozens of new journals and conferences in the subject. In part, the growth was spurred by new connections and applications to other fields, ranging from algebra to probability, from functional analysis to number theory, etc. These connections shed the boundaries between combinatorics and parts of mathematics and theoretical computer science, but at the same time led to a partial fragmentation of the field.

Approaches and Subfields of Combinatorics

Enumerative Combinatorics

Enumerative combinatorics is the most classical area of combinatorics, and concentrates on counting the number of certain combinatorial objects. Although counting the number of elements in a set is a rather broad mathematical problem, many of the problems that arise in applications have a relatively simple combinatorial description. Fibonacci numbers is the basic example of a problem in enumerative combinatorics. The twelvefold way provides a unified framework for counting permutations, combinations and partitions.

Analytic Combinatorics

Analytic combinatorics concerns the enumeration of combinatorial structures using tools from complex analysis and probability theory. In contrast with enumerative combinatorics which uses explicit combinatorial formulae and generating functions to describe the results, analytic combinatorics aims at obtaining asymptotic formulae.

Partition Theory

Partition theory studies various enumeration and asymptotic problems related to integer partitions, and is closely related to q-series, special functions and orthogonal polynomials. Originally a part of number theory and analysis, it is now considered a part of combinatorics or an independent field. It incorporates the bijective approach and various tools in analysis, analytic number theory, and has connections with statistical mechanics.

Graph Theory

Graphs are basic objects in combinatorics. The questions range from counting (e.g., the number of graphs on n vertices with k edges) to structural (e.g., which graphs contain Hamiltonian cycles) to algebraic questions (e.g., given a graph G and two numbers x and y, does the Tutte polynomial $T_G(x,y)$ have a combinatorial interpretation?). It should be noted that while there are very strong connections between graph theory and combinatorics, these two are sometimes thought of as separate subjects.

Design Theory

Design theory is a study of combinatorial designs, which are collections of subsets with certain intersection properties. Block designs are combinatorial designs of a special type. This area is one of the oldest parts of combinatorics, such as in Kirkman's schoolgirl problem proposed in 1850. The solution of the problem is a special case of a Steiner system, which systems play an important role in the classification of finite simple groups. The area has further connections to coding theory and geometric combinatorics.

Finite Geometry

Finite geometry is the study of geometric systems having only a finite number of points. Structures analogous to those found in continuous geometries (Euclidean plane, real projective space, etc.) but defined combinatorially are the main items studied. This area provides a rich source of examples for Design theory. It should not be confused with Discrete geometry (Combinatorial geometry).

Order Theory

Order theory is the study of partially ordered sets, both finite and infinite. Various examples of partial orders appear in algebra, geometry, number theory and throughout combinatorics and graph theory. Notable classes and examples of partial orders include lattices and Boolean algebras.

Matroid Theory

Matroid theory abstracts part of geometry. It studies the properties of sets (usually, finite sets) of vectors in a vector space that do not depend on the particular coefficients in a linear dependence relation. Not only the structure but also enumerative properties belong to matroid theory. Matroid theory was introduced by Hassler Whitney and studied as a part of the order theory. It is now an independent field of study with a number of connections with other parts of combinatorics.

Extremal Combinatorics

Extremal combinatorics studies extremal questions on set systems. The types of questions addressed in this case are about the largest possible graph which satisfies certain properties. For example, the largest triangle-free graph on $2n$ vertices is a complete bipartite graph $K_{n,n}$. Often it is too hard even to find the extremal answer $f(n)$ exactly and one can only give an asymptotic estimate.

Ramsey theory is another part of extremal combinatorics. It states that any sufficiently large configuration will contain some sort of order. It is an advanced generalization of the pigeonhole principle.

Probabilistic Combinatorics

In probabilistic combinatorics, the questions are of the following type: what is the probability of a certain property for a random discrete object, such as a random graph? For instance, what is the average number of triangles in a random graph? Probabilistic methods are also used to determine the existence of combinatorial objects with certain prescribed properties (for which explicit examples might be difficult to find), simply by observing that the probability of randomly selecting an object with those properties is greater than 0. This approach (often referred to as *the* probabilistic method) proved highly effective in applications to extremal combinatorics and graph theory. A closely related area is the study of finite Markov chains, especially on combinatorial objects. Here again probabilistic tools are used to estimate the mixing time.

Often associated with Paul Erdõs, who did the pioneer work on the subject, probabilistic combinatorics was traditionally viewed as a set of tools to study problems in other parts of combinatorics. However, with the growth of applications to analysis of algorithms in computer science, as well as classical probability, additive and probabilistic number theory, the area recently grew to become an independent field of combinatorics.

Algebraic Combinatorics

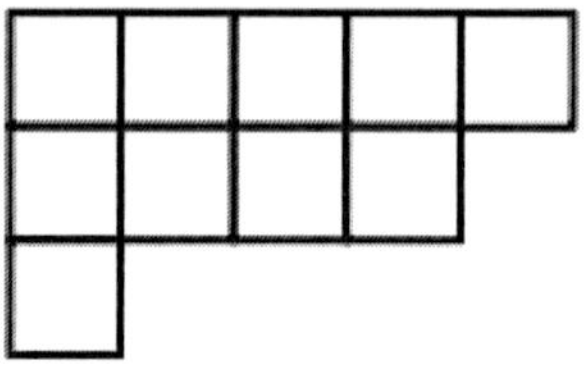

Figure: *Young diagram of a partition (5,4,1).*

Algebraic combinatorics is an area of mathematics that employs methods of abstract algebra, notably group theory and representation theory, in various combinatorial contexts and, conversely, applies combinatorial techniques to problems in algebra. Algebraic combinatorics is continuously expanding its scope, in both topics and techniques, and can be seen as the area of mathematics where the interaction of combinatorial and algebraic methods is particularly strong and significant.

Combinatorics on Words

Combinatorics on words deals with formal languages. It arose independently within several branches of mathematics, including number theory, group theory and probability. It has applications to enumerative combinatorics, fractal analysis, theoretical computer science, automata theory and linguistics. While many applications are new, the classical Chomsky–Schützenberger hierarchy of classes of formal grammars is perhaps the best known result in the field.

Geometric Combinatorics

Geometric combinatorics is related to convex and discrete geometry, in particular polyhedral combinatorics. It asks, for example, how many faces of each dimension can a convex polytope have. Metric properties of polytopes play an important role as well, e.g. the Cauchy theorem on rigidity of convex polytopes. Special polytopes are also considered, such as permutohedra, associahedra and Birkhoff polytopes. We should note that combinatorial geometry is an old fashioned name for discrete geometry.

Topological Combinatorics

Combinatorial analogs of concepts and methods in topology are used to study graph colouring, fair division, partitions, partially ordered sets, decision trees, necklace problems and discrete Morse theory. It should not be confused with combinatorial topology which is an older name for algebraic topology.

Arithmetic Combinatorics

Arithmetic combinatorics arose out of the interplay between number theory, combinatorics, ergodic theory and harmonic analysis. It is about combinatorial estimates associated with arithmetic operations (addition, subtraction, multiplication, and division). *Additive combinatorics* refers to the special case when only the operations of addition and subtraction are involved. One important technique in arithmetic combinatorics is the ergodic theory of dynamical systems.

Infinitary Combinatorics

Infinitary combinatorics, or combinatorial set theory, is an extension of ideas in combinatorics to infinite sets. It is a part of set theory, an area of mathematical logic, but uses tools and ideas from both set theory and extremal combinatorics.

Gian-Carlo Rota used the name *continuous combinatorics* to describe probability and measure theory, since there are many analogies between *counting* and *measure.*

Related Fields

Combinatorial Optimization

Combinatorial optimization is the study of optimization on discrete and combinatorial objects. It started as a part of combinatorics and graph theory, but is now viewed as a branch of applied mathematics and computer science, related to operations research, algorithm theory and computational complexity theory.

Coding Theory

Coding theory started as a part of design theory with early combinatorial constructions of error-correcting codes. The main idea of the subject is to design efficient and reliable methods of data transmission. It is now a large field of study, part of information theory.

Discrete and Computational Geometry

Discrete geometry (also called combinatorial geometry) also began a part of combinatorics, with early results on convex polytopes and kissing numbers. With the emergence of applications of discrete geometry to computational geometry, these two fields partially merged and became a separate field of study.

There remain many connections with geometric and topological combinatorics, which themselves can be viewed as outgrowths of the early discrete geometry.

Combinatorics and Dynamical Systems

Combinatorial aspects of dynamical systems is another emerging field. Here dynamical systems can be defined on combinatorial objects.

Combinatorics and Physics

There are increasing interactions between combinatorics and physics, particularly statistical physics. Examples include an exact solution of the Ising model, and a connection between the Potts model on one hand, and the chromatic and Tutte polynomials on the other hand.

Chapter 4

Probability

Probability is the measure of the likeliness that an event will occur. Probability is used to quantify an attitude of mind towards some proposition of whose truth we are not certain. The proposition of interest is usually of the form "Will a specific event occur?" The attitude of mind is of the form "How certain are we that the event will occur?" The certainty we adopt can be described in terms of a numerical measure and this number, between 0 and 1 (where 0 indicates impossibility and 1 indicates certainty), we call probability. Thus the higher the probability of an event, the more certain we are that the event will occur. A simple example would be the toss of a fair coin. Since the 2 outcomes are deemed equiprobable, the probability of "heads" equals the probability of "tails" and each probability is 1/2 or equivalently a 50% chance of either "heads" or "tails".

These concepts have been given an axiomatic mathematical formalization in probability theory, which is used widely in such areas of study as mathematics, statistics, finance, gambling, science (in particular physics), artificial intelligence/machine learning, computer science, and philosophy to, for example, draw inferences about the expected frequency of events. Probability theory is also used to describe the underlying mechanics and regularities of complex systems.

Probability Interpretations

The word probability has been used in a variety of ways since it was first applied to the mathematical study of games of chance. Does probability measure the real, physical tendency of something to occur or is it a measure of how strongly one believes it will occur, or does it draw on both these elements? In answering such questions, mathematicians interpret the probability values of probability theory.

There are two broad categories of probability interpretations which can be called "physical" and "evidential" probabilities. Physical probabilities, which are also called objective or frequency probabilities, are associated with random physical systems such as roulette wheels, rolling dice and radioactive atoms. In such systems, a given type of event (such as the dice yielding a six) tends to occur at a persistent rate, or "relative frequency", in a long run of trials. Physical probabilities either explain, or are invoked to explain, these stable frequencies. Thus talking about physical probability makes sense only when dealing with well defined random experiments. The two main kinds of theory of physical probability are frequentist accounts (such as those of Venn, Reichenbach and von Mises) and propensity accounts (such as those of Popper, Miller, Giere and Fetzer).

Evidential probability, also called Bayesian probability (or subjectivist probability), can be assigned to any statement whatsoever, even when no random process is involved, as a way to represent its subjective plausibility, or the degree to which the statement is supported by the available evidence. On most accounts, evidential probabilities are considered to be degrees of belief, defined in terms of dispositions to gamble at certain odds. The four main evidential interpretations are the classical (e.g. Laplace's) interpretation, the subjective interpretation (de Finetti and Savage), the epistemic or inductive interpretation (Ramsey, Cox) and the logical interpretation (Keynes and Carnap).

Some interpretations of probability are associated with approaches to statistical inference, including theories of estimation and hypothesis testing. The physical interpretation, for example, is taken by followers of "frequentist" statistical methods, such as R. A. Fisher, Jerzy Neyman and Egon Pearson. Statisticians of the opposing Bayesian school typically accept the existence and importance of physical probabilities, but also consider the calculation of evidential probabilities to be both valid and necessary in statistics. This article, however, focuses on the interpretations of probability rather than theories of statistical inference.

The terminology of this topic is rather confusing, in part because probabilities are studied within a variety of academic fields. The word "frequentist" is especially tricky. To philosophers it refers to a particular theory of physical probability, one that has more or less been abandoned. To scientists, on the other hand, "frequentist probability" is just another name for physical (or objective) probability. Those who promote Bayesian inference view "frequentist statistics" as an approach to statistical inference that recognises only physical probabilities. Also the word "objective", as applied to probability, sometimes means exactly what "physical" means here, but is also used of evidential probabilities that are fixed by rational constraints, such as logical and epistemic probabilities.

It is unanimously agreed that statistics depends somehow on probability. But, as to what probability is and how it is connected with statistics, there has seldom been such complete disagreement and breakdown of communication since the Tower of Babel. Doubtless, much of the disagreement is merely terminological and would disappear under sufficiently sharp analysis.

—(Savage, 1954, p 2)

Philosophy

The philosophy of probability presents problems chiefly in matters of epistemology and the uneasy interface between mathematical concepts and ordinary language as it is used by non-mathematicians. Probability theory is an established field of study in mathematics. It has its origins in correspondence discussing the mathematics of games of chance between Blaise Pascal and Pierre de Fermat in the seventeenth century, and was formalized and rendered axiomatic as a distinct branch of mathematics by Andrey Kolmogorov in the twentieth century. In its axiomatic form, mathematical statements about probability theory carry the same sort of epistemological confidence shared by other mathematical statements in the philosophy of mathematics.

The mathematical analysis originated in observations of the behaviour of game equipment such as playing cards and dice, which are designed specifically to introduce random and equalized elements; in mathematical terms, they are subjects of indifference. This is not the only way probabilistic statements are used in ordinary human language: when people say that "*it will probably rain*", they typically do not mean that the outcome of rain versus not-rain is a random factor that the odds currently favour; instead, such statements are perhaps better understood as qualifying their expectation of rain with a degree of confidence. Likewise, when it is written that "the most probable explanation" of the name of Ludlow, Massachusetts "is that it was named after Roger Ludlow", what is meant here is not that Roger Ludlow is favoured by a random factor, but rather that this is the most plausible explanation of the evidence, which admits other, less likely explanations.

Thomas Bayes attempted to provide a logic that could handle varying degrees of confidence; as such, Bayesian probability is an attempt to recast the representation of probabilistic statements as an expression of the degree of confidence by which the beliefs they express are held.

Though probability initially had somewhat mundane motivations, its modern influence and use is widespread ranging from Evidence based medicine, through Six sigma, all the way to the Probabilistically checkable proof and the String theory landscape.

A summary of some interpretations of probability				
	Classical	***Frequentist***	***Subjective***	***Propensity***
Main hypothesis	Principle of indifference	Frequency of occurrence	Degree of belief	Degree of causal connection
Conceptual basis	Hypothetical symmetry	Past data and reference class	Knowledge and intuition	Present state of system
Conceptual approach	Conjectural	Empirical	Subjective	Metaphysical
Single case possible	Yes	No	Yes	Yes
Precise	Yes	No	No	Yes
Problems	Ambiguity in principle of indifference	Reference class problem	Unchecked opinion	Disputed concept

Classical Definition

The first attempt at mathematical rigour in the field of probability, championed by Pierre-Simon Laplace, is now known as the classical definition. Developed from studies of games of chance (such as rolling dice) it states that probability is shared equally between all the possible outcomes, provided these outcomes can be deemed equally likely. (3.1)

The theory of chance consists in reducing all the events of the same kind to a certain number of cases equally possible, that is to say, to such as we may be equally undecided about in regard to their existence, and in determining the number of cases favourable to the event whose probability is sought. The ratio of this number to that of all the cases possible is the measure of this probability, which is thus simply a fraction whose numerator is the number of favourable cases and whose denominator is the number of all the cases possible.

— Pierre-Simon Laplace, *A Philosophical Essay on Probabilities*

Figure: *The classical definition of probability works well for situations with only a finite number of equally-likely outcomes.*

This can be represented mathematically as follows: If a random experiment can result in N mutually exclusive and equally likely outcomes and if N_A of these outcomes result in the occurrence of the event A, the probability of A is defined by

$$P(A) = \frac{N_A}{N}.$$

There are two clear limitations to the classical definition. Firstly, it is applicable only to situations in which there is only a 'finite' number of possible outcomes. But some important random experiments, such as tossing a coin until it rises heads, give rise to an infinite set of outcomes. And secondly, you need to determine in advance that all the possible outcomes are equally likely without relying on the notion of probability to avoid circularity—for instance, by symmetry considerations.

Frequentism

Frequentists posit that the probability of an event is its relative frequency over time, (3.4) i.e., its relative frequency of occurrence after repeating a process a large number of times under similar conditions. This is also known as aleatory probability. The events are assumed to be governed by some random physical phenomena, which are either phenomena that are predictable, in principle, with sufficient information; or phenomena which are essentially unpredictable. Examples of the first kind include tossing dice or spinning a roulette wheel; an example of the second kind is radioactive decay. In the case of tossing a fair coin, frequentists say that the probability of getting a heads is 1/2, not because there are two equally likely outcomes but because repeated series of large numbers of trials demonstrate that the empirical frequency converges to the limit 1/2 as the number of trials goes to infinity.

If we denote by n_a the number of occurrences of an event A in n trials, then if $\lim_{n\to\infty} \frac{n_a}{n} = p$ we say that $P(\mathcal{A}) = p$

The frequentist view has its own problems. It is of course impossible to actually perform an infinity of repetitions of a random experiment to determine the probability of an event. But if only a finite number of repetitions of the process are performed, different relative frequencies will appear in different series of trials. If these relative frequencies are to define the probability, the probability will be slightly different every time it is measured. But the real probability should be the same every time. If we acknowledge the fact that we only can measure a probability with some error of measurement attached, we still get into problems as the error of measurement can only be expressed as a probability, the very concept we are trying to define. This renders even the frequency definition circular.

Logical, Epistemic, and Inductive Probability

It is widely recognised that the term “probability” is sometimes used in contexts where it has nothing to do with physical randomness. Consider, for example, the claim that the extinction of the dinosaurs was probably caused by a large meteorite hitting the earth. Statements such as “Hypothesis H is probably true” have been interpreted to mean that the (presently available) empirical evidence (E, say) supports H to a high degree. This degree of support of H by E has been called the logical probability of H given E, or the epistemic probability of H given E, or the inductive probability of H given E.

The differences between these interpretations are rather small, and may seem inconsequential. One of the main points of disagreement lies in the relation between probability and belief. Logical probabilities are conceived (for example in Keynes’ Treatise on Probability) to be objective, logical relations between propositions (or sentences), and hence not to depend in any way upon belief. They are degrees of (partial) entailment, or degrees of logical consequence, not degrees of belief. (They do, nevertheless, dictate proper degrees of belief, as is discussed below.) Frank P. Ramsey, on the other hand, was skeptical about the existence of such objective logical relations and argued that (evidential) probability is “the logic of partial belief”. (p 157) In other words, Ramsey held that epistemic probabilities simply *are* degrees of rational belief, rather than being logical relations that merely *constrain* degrees of rational belief.

Another point of disagreement concerns the *uniqueness* of evidential probability, relative to a given state of knowledge. Rudolf Carnap held, for example, that logical principles always determine a unique logical probability for any statement, relative to any body of evidence. Ramsey, by contrast, thought that while degrees of belief are subject to some rational constraints (such as, but not limited to, the axioms of probability) these constraints usually do not determine a unique value. Rational people, in other words, may differ somewhat in their degrees of belief, even if they all have the same information.

Propensity

Propensity theorists think of probability as a physical propensity, or disposition, or tendency of a given type of physical situation to yield an outcome of a certain kind or to yield a long run relative frequency of such an outcome. This kind of objective probability is sometimes called 'chance'.

Propensities, or chances, are not relative frequencies, but purported causes of the observed stable relative frequencies. Propensities are invoked to explain why repeating a certain kind of experiment will generate given outcome types at persistent rates, which are known as propensities or chances. Frequentists are unable to take this approach, since relative frequencies do not exist for single tosses of a coin, but only for large ensembles or collectives. In contrast, a propensitist is able to use the law of large numbers to explain the behaviour of long-run frequencies. This law, which is a consequence of the axioms of probability, says that if (for example) a coin is tossed repeatedly many times, in such a way that its probability of landing heads is the same on each toss, and the outcomes are probabilistically independent, then the relative frequency of heads will be close to the probability of heads on each single toss. This law allows that stable long-run frequencies are a manifestation of invariant *single-case* probabilities. In addition to explaining the emergence of stable relative frequencies, the idea of propensity is motivated by the desire to make sense of single-case probability attributions in quantum mechanics, such as the probability of decay of a particular atom at a particular time.

The main challenge facing propensity theories is to say exactly what propensity means. (And then, of course, to show that propensity thus defined has the required properties.) At present, unfortunately, none of the well-recognised accounts of propensity comes close to meeting this challenge.

A propensity theory of probability was given by Charles Sanders Peirce. A later propensity theory was proposed by philosopher Karl

Popper, who had only slight acquaintance with the writings of C. S. Peirce, however. Popper noted that the outcome of a physical experiment is produced by a certain set of "generating conditions". When we repeat an experiment, as the saying goes, we really perform another experiment with a (more or less) similar set of generating conditions. To say that a set of generating conditions has propensity p of producing the outcome E means that those exact conditions, if repeated indefinitely, would produce an outcome sequence in which E occurred with limiting relative frequency p. For Popper then, a deterministic experiment would have propensity 0 or 1 for each outcome, since those generating conditions would have same outcome on each trial. In other words, non-trivial propensities (those that differ from 0 and 1) only exist for genuinely in deterministic experiments.

A number of other philosophers, including David Miller and Donald A. Gillies, have proposed propensity theories somewhat similar to Popper's.

Other propensity theorists (e.g. Ronald Giere) do not explicitly define propensities at all, but rather see propensity as defined by the theoretical role it plays in science. They argue, for example, that physical magnitudes such as electrical charge cannot be explicitly defined either, in terms of more basic things, but only in terms of what they do (such as attracting and repelling other electrical charges). In a similar way, propensity is whatever fills the various roles that physical probability plays in science.

What roles does physical probability play in science? What are its properties? One central property of chance is that, when known, it constrains rational belief to take the same numerical value. David Lewis called this the *Principal Principle*, (3.3 & 3.5) a term that philosophers have mostly adopted. For example, suppose you are certain that a particular biased coin has propensity 0.32 to land heads every time it is tossed. What is then the correct price for a gamble that pays $1 if the coin lands heads, and nothing otherwise? According to the Principal Principle, the fair price is 32 cents.

Subjectivism

Subjectivists, also known as Bayesians or followers of epistemic probability, give the notion of probability a subjective status by regarding it as a measure of the 'degree of belief' of the individual assessing the uncertainty of a particular situation. Epistemic or subjective probability is sometimes called credence, as opposed to the term chance for a propensity probability.

Some examples of epistemic probability are to assign a probability to the proposition that a proposed law of physics is true, and to determine how probable it is that a suspect committed a crime, based on the evidence presented.

Gambling odds don't reflect the bookies' belief in a likely winner, so much as the other bettors' belief, because the bettors are actually betting against one another. The odds are set based on how many people have bet on a possible winner, so that even if the high odds players always win, the bookies will always make their percentages anyway.

The use of Bayesian probability raises the philosophical debate as to whether it can contribute valid justifications of belief.

Bayesians point to the work of Ramsey (p 182) and de Finetti (p 103) as proving that subjective beliefs must follow the laws of probability if they are to be coherent. Evidence casts doubt on human coherence.

The use of Bayesian probability involves specifying a prior probability. This may be obtained from consideration of whether the required prior probability is greater or lesser than a reference probability associated with an urn model or a thought experiment. The issue is that for a given problem, multiple thought experiments could apply, and choosing one is a matter of judgement: different people may assign different prior probabilities, known as the reference class problem. The "sunrise problem" provides an example.

Prediction

An alternative account of probability emphasizes the role of *prediction* – predicting future observations on the basis of past observations, not on unobservable parameters. In its modern form, it is mainly in the Bayesian vein. This was the main function of probability before the 20th century, but fell out of favour compared to the parametric approach, which modelled phenomena as a physical system that was observed with error, such as in celestial mechanics.

The modern predictive approach was pioneered by Bruno de Finetti, with the central idea of exchangeability – that future observations should behave like past observations. This view came to the attention of the Anglophone world with the 1974 translation of de Finetti's book, and has since been propounded by such statisticians as Seymour Geisser.

Axiomatic Probability

The mathematics of probability can be developed on an entirely axiomatic basis that is independent of any interpretation.

Probability Theory

Probability theory is the branch of mathematics concerned with probability, the analysis of random phenomena. The central objects of probability theory are random variables, stochastic processes, and events: mathematical abstractions of non-deterministic events or measured quantities that may either be single occurrences or evolve over time in an apparently random fashion. If an individual coin toss or the roll of dice is considered to be a random event, then if repeated many times the sequence of random events will exhibit certain patterns, which can be studied and predicted. Two representative mathematical results describing such patterns are the law of large numbers and the central limit theorem.

As a mathematical foundation for statistics, probability theory is essential to many human activities that involve quantitative analysis of large sets of data. Methods of probability theory also apply to descriptions of complex systems given only partial knowledge of their state, as in statistical mechanics. A great discovery of twentieth century physics was the probabilistic nature of physical phenomena at atomic scales, described in quantum mechanics.

History

The mathematical theory of probability has its roots in attempts to analyze games of chance by Gerolamo Cardano in the sixteenth century, and by Pierre de Fermat and Blaise Pascal in the seventeenth century (for example the "problem of points"). Christiaan Huygens published a book on the subject in 1657 and in the 19th century a big work was done by Laplace in what can be considered today as the classic interpretation. Initially, probability theory mainly considered discrete events, and its methods were mainly combinatorial. Eventually, analytical considerations compelled the incorporation of continuous variables into the theory.

This culminated in modern probability theory, on foundations laid by Andrey Nikolaevich Kolmogorov. Kolmogorov combined the notion of sample space, introduced by Richard von Mises, and measure theory and presented his axiom system for probability theory in 1933. Fairly quickly this became the mostly undisputed axiomatic basis for modern probability theory but alternatives exist, in particular the adoption of finite rather than countable additivity by Bruno de Finetti.

Treatment

Most introductions to probability theory treat discrete probability distributions and continuous probability distributions separately. The more mathematically advanced measure theory based treatment of probability covers both the discrete, the continuous, any mix of these two and more.

Motivation

Consider an experiment that can produce a number of outcomes. The set of all outcomes is called the *sample space* of the experiment. The *power set* of the sample space is formed by considering all different collections of possible results. For example, rolling an honest die produces one of six possible results. One collection of possible results corresponds to getting an odd number. Thus, the subset {1,3,5} is an element of the power set of the sample space of die rolls. These collections are called *events*. In this case, {1,3,5} is the event that the die falls on some odd number. If the results that actually occur fall in a given event, that event is said to have occurred.

Probability is a way of assigning every "event" a value between zero and one, with the requirement that the event made up of all possible results (in our example, the event {1,2,3,4,5,6}) be assigned a value of one. To qualify as a probability distribution, the assignment of values must satisfy the requirement that if you look at a collection of mutually exclusive events (events that contain no common results, e.g., the events {1,6}, {3}, and {2,4} are all mutually exclusive), the probability that one of the events will occur is given by the sum of the probabilities of the individual events.

The probability that any one of the events {1,6}, {3}, or {2,4} will occur is 5/6. This is the same as saying that the probability of event {1,2,3,4,6} is 5/6. This event encompasses the possibility of any number except five being rolled. The mutually exclusive event {5} has a probability of 1/6, and the event {1,2,3,4,5,6} has a probability of 1, that is, absolute certainty.

Discrete Probability Distributions

Discrete probability theory deals with events that occur in countable sample spaces.

Examples: Throwing dice, experiments with decks of cards, random walk, and tossing coins

Classical definition: Initially the probability of an event to occur was defined as number of cases favourable for the event, over the number of total outcomes possible in an equiprobable sample space. For example, if the event is "occurrence of an even number when a die is rolled", the probability is given by $\frac{3}{6} = \frac{1}{2}$, since 3 faces out of the 6 have even numbers and each face has the same probability of appearing.

Modern definition: The modern definition starts with a finite or countable set called the sample space, which relates to the set of all *possible outcomes* in classical sense, denoted by Ω. It is then assumed that for each element $x \in \Omega$, an intrinsic "probability" value $f(x)$ is attached, which satisfies the following properties:

1. $f(x) \in [0,1]$ for all $x \in \Omega$;
2. $\sum_{x \in \Omega} f(x) = 1.$

That is, the probability function $f(x)$ lies between zero and one for every value of x in the sample space Ω, and the sum of $f(x)$ over all values x in the sample space Ω is equal to 1. An event is defined as any subset E of the sample space Ω. The probability of the event E is defined as

$$P(E) = \sum_{x \in E} f(x).$$

So, the probability of the entire sample space is 1, and the probability of the null event is 0.

The function $f(x)$ mapping a point in the sample space to the "probability" value is called a probability mass function abbreviated as pmf. The modern definition does not try to answer how probability mass functions are obtained; instead it builds a theory that assumes their existence.

Continuous Probability Distributions

Continuous probability theory deals with events that occur in a continuous sample space.

Classical definition: The classical definition breaks down when confronted with the continuous case.

Modern definition: If the outcome space of a random variable X is the set of real numbers ($\mathbb{R}$) F or a subset thereof, then a function called the cumulative distribution function (or cdf) exists, defined by $F(x) = P(X \leq x)$. That is, $F(x)$ returns the probability that X will be less than or equal to x.

The cdf necessarily satisfies the following properties.

1. F is a monotonically non-decreasing, right-continuous function;
2. $\lim_{x \to -\infty} F(x) = 0;$
3. $\lim_{x \to \infty} F(x) = 1.$

If F is absolutely continuous, i.e., its derivative exists and integrating the derivative gives us the cdf back again, then the random variable X is said to have a probability density function or pdf or simply density $f(x) = \frac{dF(x)}{dx}$.

For a set , $E \subseteq \mathbb{R}$ the probability of the random variable X being in E is

$$P(X \in E) = \int_{x \in E} dF(x).$$

In case the probability density function exists, this can be written as:

$$P(X \in E) = \int_{x \in E} f(x)dx.$$

Whereas the *pdf* exists only for continuous random variables, the *cdf* exists for all random variables (including discrete random variables) that take values in

These concepts can be generalized for multidimensional cases on $\mathbb{R}^n$ and other continuous sample spaces.

Measure-Theoretic Probability Theory

The *raison d'être* of the measure-theoretic treatment of probability is that it unifies the discrete and the continuous cases, and makes the difference a question of which measure is used. Furthermore, it covers distributions that are neither discrete nor continuous nor mixtures of the two.

An example of such distributions could be a mix of discrete and continuous distributions—for example, a random variable that is 0 with probability 1/2, and takes a random value from a normal distribution with probability 1/2. It can still be studied to some extent by considering it to have a pdf of $(\delta[x] + \varphi(x))/2$, where $\delta[x]$ is the Dirac delta function.

Other distributions may not even be a mix, for example, the Cantor distribution has no positive probability for any single point, neither does it have a density. The modern approach to probability theory solves these problems using measure theory to define the probability space:

Given any set Ω , (also called sample space) and a σ-algebra $\mathcal{F}$ on it, a measure P defined on $\mathcal{F}$ is called a probability measure if $P(\Omega) = 1$.

If $\mathcal{F}$ is the Borel σ-algebra on the set of real numbers, then there is a unique probability measure on $\mathcal{F}$ for any cdf, and vice versa. The measure corresponding to a cdf is said to be induced by the cdf. This measure coincides with the pmf for discrete variables and pdf for continuous variables, making the measure-theoretic approach free of fallacies.

The *probability* of a set E in the σ-algebra $\mathcal{F}$ is defined as

$$P(E) = \int_{\omega \in E} \mu_F(d\omega)$$

where the integration is with respect to the measure μ_F induced by

Along with providing better understanding and unification of discrete and continuous probabilities, measure-theoretic treatment also allows us to work on probabilities outside $\mathbb{R}^n$, as in the theory of stochastic processes. For example to study Brownian motion, probability is defined on a space of functions.

Probability Distributions

Certain random variables occur very often in probability theory because they well describe many natural or physical processes. Their distributions therefore have gained *special importance* in probability theory. Some fundamental *discrete distributions* are the discrete uniform, Bernoulli, binomial, negative binomial, Poisson and geometric distributions. Important *continuous distributions* include the continuous uniform, normal, exponential, gamma and beta distributions.

Convergence of Random Variables

In probability theory, there are several notions of convergence for random variables. They are listed below in the order of strength, i.e., any subsequent notion of convergence in the list implies convergence according to all of the preceding notions.

Weak convergence: A sequence of random variables $X_1, X_2, \ldots,$ converges weakly to the random variable X if their respective cumulative *distribution functions* $F_1, F_2, \ldots$ converge to the cumulative distribution function F of X, wherever F is continuous. Weak convergence is also called convergence in distribution.

Most common shorthand notation: $X_n \xrightarrow{\mathcal{D}} X$.

Convergence in probability: The sequence of random variables $X_1, X_2, \ldots$ is said to converge towards the random variable X in probability if $\lim_{n\to\infty} P(|X_n - X| \geq \varepsilon) = 0$ for every ε > 0.

Most common shorthand notation: $X_n \xrightarrow{P} X$.

Strong convergence: The sequence of random variables $X_1, X_2, \ldots$ is said to converge towards the random variable X strongly if $P(\lim_{n \to \infty} X_n = X) = 1$. Strong convergence is also known as almost sure convergence.

Most Common Shorthand Notation: As the names indicate, weak convergence is weaker than strong convergence. In fact, strong convergence implies convergence in probability, and convergence in probability implies weak convergence. The reverse statements are not always true.

Law of Large Numbers

Common intuition suggests that if a fair coin is tossed many times, then *roughly* half of the time it will turn up *heads*, and the other half it will turn up *tails*. Furthermore, the more often the coin is tossed, the more likely it should be that the ratio of the number of *heads* to the number of *tails* will approach unity. Modern probability provides a formal version of this intuitive idea, known as the law of large numbers. This law is remarkable because it is not assumed in the foundations of probability theory, but instead emerges from these foundations as a theorem. Since it links theoretically derived probabilities to their actual frequency of occurrence in the real world, the law of large numbers is considered as a pillar in the history of statistical theory and has had widespread influence.

The law of large numbers (LLN) states that the sample average

$$\overline{X}_n = \frac{1}{n}\sum_{k=1}^{n} X_k$$

of a sequence of independent and identically distributed random variables X_k converges towards their common expectation μ, provided that the expectation of $|X_k|$ is finite.

It is in the different forms of convergence of random variables that separates the *weak* and the *strong* law of large numbers

$$\text{Weak law:} \quad \overline{X}_n \xrightarrow{P} \mu \quad \text{for } n \to \infty$$

$$\text{Strong law:} \quad \overline{X}_n \xrightarrow{\text{a.s.}} \mu \quad \text{for } n \to \infty.$$

It follows from the LLN that if an event of probability p is observed repeatedly during independent experiments, the ratio of the observed frequency of that event to the total number of repetitions converges towards p.

For example, if $Y_1, Y_2, \ldots$ are independent Bernoulli random variables taking values 1 with probability p and 0 with probability 1-p, then $E(Y_i) = p$ for all i, so that $\bar{Y}_n$ converges to p almost surely.

Central Limit Theorem

"The central limit theorem (CLT) is one of the great results of mathematics." (Chapter 18 in) It explains the ubiquitous occurrence of the normal distribution in nature.

The theorem states that the average of many independent and identically distributed random variables with finite variance tends towards a normal distribution *irrespective* of the distribution followed by the original random variables. Formally, let $X_1, X_2, \ldots$ be independent random variables with mean μ and variance $\sigma^2 > 0$. Then the sequence of random variables

$$Z_n = \frac{\sum_{i=1}^{n}(X_i - \mu)}{\sigma\sqrt{n}}$$

converges in distribution to a standard normal random variable.

Notice that for some classes of random variables the classic central limit theorem works rather fast (Berry–Esseen theorem), for example the distributions with finite first, second and third moment from the exponential family, on the other hand for some random variables of the heavy tail and fat tail variety, it works very slow or may not work at all: in such cases one may use the Generalized Central Limit Theorem (GCLT).

Law of Total Probability

In probability theory, the law (or formula) of total probability is a fundamental rule relating marginal probabilities to conditional probabilities. It expresses the total probability of an outcome which can be realised via several distinct events - hence the name.

Statement

The law of total probability is the proposition that if $\{B_n : n = 1, 2, 3, \ldots\}$ is a finite or countably infinite partition of a sample space (in other words, a set of pairwise disjoint events whose union is the entire sample space) and each event B_n is measurable, then for any event A of the same probability space:

$$\Pr(A) = \sum_n \Pr(A \cap B_n)$$

or, alternatively,

$$\Pr(A) = \sum_n \Pr(A \mid B_n)\Pr(B_n),$$

where, for any n for which $\Pr(B_n) = 0$ these terms are simply omitted from the summation, because $\Pr(A \mid B_n)$ is finite.

The summation can be interpreted as a weighted average, and consequently the marginal probability, $\Pr(A)$, is sometimes called "average probability"; "overall probability" is sometimes used in less formal writings.

The law of total probability can also be stated for conditional probabilities. Taking the B_n as above, and assuming C is an event independent with any of the :

$$\Pr(A \mid C) = \sum_n \Pr(A \mid C \cap B_n)\Pr(B_n \mid C) = \sum_n \Pr(A \mid C \cap B_n)\Pr(B_n)$$

Informal Formulation

The above mathematical statement might be interpreted as follows: *given an outcome A, with known conditional probabilities given any of the B_n events, each with a known probability itself, what is the total probability that A will happen?*. The answer to this question is given by $\Pr(A)$.

Example

Suppose that two factories supply light bulbs to the market. Factory *X*'s bulbs work for over 5000 hours in 99% of cases, whereas factory *Y*'s bulbs work for over 5000 hours in 95% of cases. It is known that factory *X* supplies 60% of the total bulbs available. What is the chance that a purchased bulb will work for longer than 5000 hours?

Applying the law of total probability, we have:

$$\Pr(A) = \Pr(A \mid B_1)\cdot\Pr(B_1) + \Pr(A \mid B_2)\cdot\Pr(B_2) = \frac{99}{100}\cdot\frac{6}{10} + \frac{95}{100}\cdot\frac{4}{10} = \frac{594+380}{1000} = \frac{974}{1000}$$

where

- $\Pr(B_1) = \frac{6}{10}$ is the probability that the purchased bulb was manufactured by factory *X*;

- $\Pr(B_2) = \frac{4}{10}$ is the probability that the purchased bulb was manufactured by factory *Y*;
- $\Pr(A \mid B_1) = \frac{99}{100}$ is the probability that a bulb manufactured by *X* will work for over 5000h;
- $\Pr(A \mid B_2) = \frac{95}{100}$ is the probability that a bulb manufactured by *Y* will work for over 5000h.

Thus each purchased light bulb has a 97.4% chance to work for more than 5000 hours.

Applications

One common application of the law is where the events coincide with a discrete random variable *X* taking each value in its range, i.e. B_n is the event $X = x_n$. It follows that the probability of an event *A* is equal to the expected value of the conditional probabilities of *A* given $X = x_n$. That is,

$$\Pr(A) = \sum_n \Pr(A \mid X = x_n)\Pr(X = x_n) = \mathrm{E}[\Pr(A \mid X)],$$

where $\Pr(A \mid X)$ is the conditional probability of *A* given the value of the random variable *X*. This conditional probability is a random variable in its own right, whose value depends on that of *X*. The conditional probability $\Pr(A \mid X = \mathrm{x})$ is simply a conditional probability given an event, $[X = x]$. It is a function of *x*, say $g(x) = \Pr(A \mid X = x)$. Then the conditional probability $\Pr(A \mid X)$ is $g(X)$, hence itself a random variable. This version of the law of total probability says that the expected value of this random variable is the same as $\Pr(A)$.

This result can be generalized to continuous random variables (via continuous conditional density), and the expression becomes

$$\Pr(A) = \mathrm{E}[\Pr(A \mid \mathcal{F}_X)],$$

where $\mathcal{F}_X$ denotes the sigma-algebra generated by the random variable *X*.

Other Names

The term *law of total probability* is sometimes taken to mean the law of alternatives, which is a special case of the law of total probability applying to discrete random variables. One author even uses the terminology "continuous law of alternatives" in the continuous case.

This result is given by Grimmett and Welsh as the partition theorem, a name that they also give to the related law of total expectation.

Mathematical Treatment

Consider an experiment that can produce a number of results. The collection of all results is called the sample space of the experiment. The power set of the sample space is formed by considering all different collections of possible results. For example, rolling a dice can produce six possible results. One collection of possible results gives an odd number on the dice. Thus, the subset {1,3,5} is an element of the power set of the sample space of dice rolls. These collections are called "events." In this case, {1,3,5} is the event that the dice falls on some odd number. If the results that actually occur fall in a given event, the event is said to have occurred.

A probability is a way of assigning every event a value between zero and one, with the requirement that the event made up of all possible results (in our example, the event {1,2,3,4,5,6}) is assigned a value of one. To qualify as a probability, the assignment of values must satisfy the requirement that if you look at a collection of mutually exclusive events (events with no common results, e.g., the events {1,6}, {3}, and {2,4} are all mutually exclusive), the probability that at least one of the events will occur is given by the sum of the probabilities of all the individual events.

The probability of an event A is written as $P(A)$, $p(A)$ or $\Pr(A)$. This mathematical definition of probability can extend to infinite sample spaces, and even uncountable sample spaces, using the concept of a measure.

The *opposite* or *complement* of an event A is the event [not A] (that is, the event of A not occurring); its probability is given by $P(\text{not } A) = 1 - P(A)$. As an example, the chance of not rolling a six on a six-sided die is $1 - (\text{chance of rolling a six}) = 1 - \frac{1}{6} = \frac{5}{6}$.

If two events A and B occur on a single performance of an experiment, this is called the intersection or joint probability of A and B, denoted as $P(A \cap B)$.

Independent Events

If two events, A and B are independent then the joint probability is

$$P(A \text{ and } B) = P(A \cap B) = P(A)P(B),$$

for example, if two coins are flipped the chance of both being heads is $\frac{1}{2} \times \frac{1}{2} = \frac{1}{4}$.

Mutually Exclusive Events

If either event A or event B or both events occur on a single performance of an experiment this is called the union of the events A and B denoted as $P(A \cup B)$. If two events are mutually exclusive then the probability of either occurring is

$$P(A \text{ or } B) = P(A \cup B) = P(A) + P(B).$$

For example, the chance of rolling a 1 or 2 on a six-sided die is $P(1 \text{ or } 2) = P(1) + P(2) = \frac{1}{6} + \frac{1}{6} = \frac{1}{3}$.

Not Mutually Exclusive Events

If the events are not mutually exclusive then

$$P(A \text{ or } B) = P(A) + P(B) - P(A \text{ and } B).$$

For example, when drawing a single card at random from a regular deck of cards, the chance of getting a heart or a face card (J,Q,K) (or one that is both) is $\frac{13}{52} + \frac{12}{52} - \frac{3}{52} = \frac{11}{26}$, because of the 52 cards of a deck 13 are hearts, 12 are face cards, and 3 are both: here the possibilities included in the "3 that are both" are included in each of the "13 hearts" and the "12 face cards" but should only be counted once.

Conditional Probability

Conditional probability is the probability of some event *A*, given the occurrence of some other event *B*. Conditional probability is written $P(A \mid B)$, and is read "the probability of *A*, given *B*". It is defined by

$$P(A \mid B) = \frac{P(A \cap B)}{P(B)}.$$

If $P(B) = 0$ then $P(A \mid B)$ is formally undefined by this expression. However, it is possible to define a conditional probability for some zero-probability events using a σ-algebra of such events (such as those arising from a continuous random variable). For example, in a bag of 2 red balls and 2 blue balls (4 balls in total), the probability of taking a red ball is $1/2$; however, when taking a second ball, the probability of it being either a red ball or a blue ball depends on the ball previously taken, such as, if a red ball was taken, the probability of picking a red ball again would be $1/3$ since only 1 red and 2 blue balls would have been remaining.

Inverse Probability

In probability theory and applications, Bayes' rule relates the odds of event A_1 to event A_2, before (prior to) and after (posterior to) conditioning on another event B. The odds A_1 on to event A_2 is simply the ratio of the probabilities of the two events. When arbitrarily many events are of A interest, not just two, the rule can be rephrased as posterior is proportional to prior times likelihood, $P(A \mid B) \propto P(A)P(B \mid A)$ where the proportionality symbol means that the left hand side is proportional to (i.e., equals a constant times) the right hand side as A varies, for fixed or given B (Lee, 2012; Bertsch McGrayne, 2012). In this form it goes back to Laplace (1774) and to Cournot (1843).

Summary of Probabilities

Summary of probabilities	
Event	***Probability***
A	$P(A) \in [0,1]$
not A	$P(A^c) = 1 - P(A)$
A or B	$P(A \cup B) = P(A) + P(B) - P(A \cap B)$ $P(A \cup B) = P(A) + P(B)$ if A and B are mutually exclusive
A and B	$P(A \cap B) = P(A \mid B)P(B) = P(B \mid A)P(A)$ $P(A \cap B) = P(A)P(B)$ if A and B are independent
A given B	$P(A \mid B) = \frac{P(A \cap B)}{P(B)} = \frac{P(B \mid A)P(A)}{P(B)}$

Relation to Randomness

In a deterministic universe, based on Newtonian concepts, there would be no probability if all conditions were known (Laplace's demon), (but there are situations in which sensitivity to initial conditions exceeds our ability to measure them, i.e. know them). In the case of a roulette wheel, if the force of the hand and the period of that force are known, the number on which the ball will stop would be a certainty (though as a practical matter, this would likely be true only of a roulette wheel that had not been exactly levelled — as Thomas A. Bass' Newtonian Casino revealed). Of course, this also assumes knowledge of inertia and friction of the wheel, weight, smoothness and roundness of the ball, variations in hand speed during the turning and so forth. A probabilistic description can thus be more useful than Newtonian mechanics for analyzing the pattern of outcomes of repeated rolls of a roulette wheel. Physicists face the same situation in kinetic theory of gases, where the system, while deterministic *in principle*, is so complex (with the number of molecules typically the order of magnitude of Avogadro constant $6.02 \cdot 10^{23}$) that only a statistical description of its properties is feasible.

Probability theory is required to describe quantum phenomena. A revolutionary discovery of early 20th century physics was the random character of all physical processes that occur at sub-atomic scales and are governed by the laws of quantum mechanics. The objective wave function evolves deterministically but, according to the Copenhagen interpretation, it deals with probabilities of observing, the outcome being explained by a wave function collapse when an observation is made. However, the loss of determinism for the sake of instrumentalism did not meet with universal approval. Albert Einstein famously remarked in a letter to Max Born: "I am convinced that God does not play dice". Like Einstein, Erwin Schrödinger, who discovered the wave function, believed quantum mechanics is a statistical approximation of an underlying deterministic reality. In modern interpretations, quantum decoherence accounts for subjectively probabilistic behaviour.

Probability Space

In probability theory, a probability space or a probability triple is a mathematical construct that models a real-world process (or "experiment") consisting of states that occur randomly. A probability space is constructed with a specific kind of situation or experiment in mind. One proposes that each time a situation of that kind arises, the set of possible outcomes is the same and the probabilities are also the same.

A probability space consists of three parts:

1. A sample space, Ω, which is the set of all possible outcomes.
2. A set of events $\mathcal{F}$, where each event is a set containing zero or more outcomes.
3. The assignment of probabilities to the events; that is, a function P from events to probabilities.

An outcome is the result of a single execution of the model. Since individual outcomes might be of little practical use, more complex *events* are used to characterize groups of outcomes. The collection of all such events is a *σ-algebra* . Finally, there is a need to specify each event's likelihood of happening. This is done using the *probability measure* function, P.

Once the probability space is established, it is assumed that "nature" makes its move and selects a single outcome, Ω, from the sample space Ω. All the events in that contain the selected outcome ω (recall that each event is a subset of Ω) are said to "have occurred". The selection performed by nature is done in such a way that if the experiment were to be repeated an infinite number of times, the relative frequencies of occurrence of each of the events would coincide with the probabilities prescribed by the function P.

The Russian mathematician Andrey Kolmogorov introduced the notion of probability space, together with other axioms of probability, in the 1930s. Nowadays alternative approaches for axiomatization of probability theory exist.

This article is concerned with the mathematics of manipulating probabilities. The article probability interpretations outlines several alternative views of what "probability" means and how it should be interpreted. In addition, there have been attempts to construct theories for quantities that are notionally similar to probabilities but do not obey all their rules.

Introduction

A probability space is a mathematical triplet $(\Omega, \mathcal{F}, P)$ that presents a model for a particular class of real-world situations. As with other models, its author ultimately defines which elements Ω, , and P will contain.

- The sample space Ω is a set of outcomes. An outcome is the result of a single execution of the model. Outcomes may be states of nature, possibilities, experimental results and the like. Every instance of the real-world situation (or run of the experiment) must produce exactly one outcome. If outcomes of different runs of an experiment differ in any way that matters, they are distinct outcomes. Which differences matter depends on the kind of analysis we want to do: This leads to different choices of sample space.
- The σ-algebra is a collection of all the events (not necessarily elementary) we would like to consider. Here, an "event" is a set of zero or more outcomes, i.e., a subset of the sample space. An event is considered to have "happened" during an experiment when the outcome of the latter is an element of the event. Since the same outcome may be a member of many events, it is possible for many events to have happened given a single outcome. For example, when the trial consists of throwing two dice, the set of all outcomes with a sum of 7 pips may constitute an event, whereas outcomes with an odd number of pips may constitute another event. If the outcome is the element of the elementary event of two pips on the first die and five on the second, then both of the events, "7 pips" and "odd number of pips", are said to have happened.

- The probability measure P is a function returning an event's probability. A probability is a real number between zero (impossible events have probability zero, though probability-zero events are not necessarily impossible) and one (the event happens almost surely, with total certainty). Thus P is a function $P: \mathcal{F} \rightarrow [0,1]$. The probability measure function must satisfy two simple requirements: First, the probability of a union of two (or countably many) disjoint events must be equal to the sum of probabilities of each of these events. For example, if two events are *Heads* and *Tails*, then the probability of *Heads-or-Tails* must be equal to the sum of probabilities for *Heads* and *Tails*). Second, the probability of the sample space Ω must be equal to 1 (which accounts for the fact that, given an execution of the model, some outcome must occur). In the previous example the probability of the set of outcomes *P*(*Heads*,*Tails*) must be equal to one, because it is entirely certain that the outcome will be either *Heads* or *Tails* (the model neglects any other possibility).

Not every subset of the sample space Ω must necessarily be considered an event: some of the subsets are simply not of interest, others cannot be "measured". This is not so obvious in a case like a coin toss. In a different example, one could consider javelin throw lengths, where the events typically are intervals like "between 60 and 65 metres" and unions of such intervals, but not sets like the "irrational numbers between 60 and 65 metres"

Definition

In short, a probability space is a measure space such that the measure of the whole space is equal to one.

The expanded definition is following: a probability space is a triple $(\Omega, \mathcal{F}, P)$ consisting of:

- the sample space Ω — an arbitrary non-empty set,
- the σ-algebra $\mathcal{F} \subseteq 2^{\Omega}$ (also called σ-field) — a set of subsets of Ω, called events, such that:
  - o contains the sample space: ,
  - o $\mathcal{F}$ is closed under complements: if $A \in \mathcal{F}$, then also $(\Omega \setminus A) \in \mathcal{F}$,
  - o $\mathcal{F}$ is closed under countable unions: if A_i" for i=1,2,..., then also $(\cup_i A_i) \in \mathcal{F}$

  - The corollary from the previous two properties and De Morgan's law is that $\mathcal{F}$ is also closed under countable intersections: if $A_i \in \mathcal{F}$ for $i = 1, 2, \ldots,$ then also $(\cap_i A_i) \in \mathcal{F}$
- the probability measure P: $\rightarrow[0,1]$ — a function on such that:
  - o P is countably additive: if $\{A_i\} \subseteq$ is a countable collection of pairwise disjoint sets, then $P(\sqcup A_i) = \sum P(A_i)$, where "$\sqcup$" denotes the disjoint union,
  - o the measure of entire sample space is equal to one: $P(\Omega) = 1$.

Discrete Case

Discrete probability theory needs only at most countable sample spaces Ω. Probabilities can be ascribed to points of Ω by the probability mass function p: $\Omega \rightarrow [0,1]$ such that $\sum_{\omega \in \Omega} p(\omega) = 1$. All subsets of Ω can be treated as events (thus, $\mathcal{F} = 2^{\Omega}$ is the power set). The probability measure takes the simple form

$$(*) \qquad P(A) = \sum_{\omega \in A} p(\omega) \quad \text{for all } A \subseteq \Omega.$$

The greatest σ-algebra $\mathcal{F} = 2^{\Omega}$ describes the complete information. In general, a σ-algebra $\mathcal{F} \subseteq 2^{\Omega}$ corresponds to a finite or countable partition $\Omega = B_1 \sqcup B_2 \sqcup \ldots,$ the general form of an event $A \in$ being $A = B_{k1} \sqcup B_{k2} \sqcup \ldots$ (here $\sqcup$ means the disjoint union.)

The case $p(\omega) = 0$ is permitted by the definition, but rarely used, since such ω can safely be excluded from the sample space.

General Case

If Ω is uncountable, still, it may happen that $p(\omega) \neq 0$ for some ω; such ω are called atoms. They are an at most countable (maybe empty) set, whose probability is the sum of probabilities of all atoms. If this sum is equal to 1 then all other points can safely be excluded from the sample space, returning us to the discrete case. Otherwise, if the sum of probabilities of all atoms is less than 1 (maybe 0), then the probability space decomposes into a discrete (atomic) part (maybe empty) and a non-atomic part.

Non-Atomic Case

If $p(\omega) = 0$ for all $\omega \in \Omega$ (in this case, Ω must be uncountable, because otherwise $P(\Omega)=1$ could not be satisfied), then equation (*)

fails: the probability of a set is not the sum over its elements, as summation is only defined for countable amount of elements. This makes the probability space theory much more technical. A formulation stronger than summation, measure theory is applicable. Initially the probabilities are ascribed to some "generator" sets.

Then a limiting procedure allows assigning probabilities to sets that are limits of sequences of generator sets, or limits of limits, and so on. All these sets are the σ-algebra $\mathcal{F}$. Sets belonging to $\mathcal{F}$ are called measurable. In general they are much more complicated than generator sets, but much better than non-measurable sets.

Complete Probability Space

A probability space $(\Omega, \mathcal{F}, P)$ is said to be a complete probability space if for all $B \in \mathcal{F}$ with $P(B) = 0$ and all $A \subset B$ one has $A \in \mathcal{F}$. Often, the study of probability spaces is restricted to complete probability spaces.

Examples

Discrete Examples

Example 1

If the experiment consists of just one flip of a perfect coin, then the outcomes are either heads or tails: $\Omega = \{H, T\}$. The σ-algebra $\mathcal{F} = 2^{\Omega}$ contains $2^2 = 4$ events, namely: {H} – "heads", {T} – "tails", – "neither heads nor tails", and {H,T} – "either heads or tails". So, = {, {H}, {T}, {H,T. There is a fifty percent chance of tossing heads, and fifty percent for tails. Thus the probability measure in this example is $P() = 0$, $P(\{H\}) = 0.5$, $P(\{T\}) = 0.5$, $P(\{H,T\}) = 1$.

Example 2

The fair coin is tossed three times. There are 8 possible outcomes: Ω = {HHH, HHT, HTH, HTT, THH, THT, TTH, TTT} (here "HTH" for example means that first time the coin landed heads, the second time tails, and the last time heads again). The complete information is described by the σ-algebra $\mathcal{F} = 2^{\Omega}$ of $2^8 = 256$ events, where each of the events is a subset of Ω.

Alice knows the outcome of the second toss only. Thus her incomplete information is described by the partition $\Omega = A_1 \sqcup A_2$ = {HHH, HHT, THH, THT} $\sqcup$ {HTH, HTT, TTH, TTT}, and the corresponding σ-algebra $_{\text{Alice}} = \{, A_1, A_2, \Omega\}$. Brian knows only the total number of tails. His partition contains four parts: $\Omega = B_0 \sqcup B_1 \sqcup B_2 \sqcup$

B_3 = {HHH} $\sqcup$ {HHT, HTH, THH} $\sqcup$ {TTH, THT, HTT} $\sqcup$ {TTT}; accordingly, his σ-algebra $_{\text{Brian}}$ contains $2^4 = 16$ events.

The two σ-algebras are incomparable: neither $_{\text{Alice}} \subseteq {}_{\text{Brian}}$ nor $_{\text{Brian}} \subseteq {}_{\text{Alice}}$; both are sub-σ-algebras of 2^Ω.

Example 3

If 100 voters are to be drawn randomly from among all voters in California and asked whom they will vote for governor, then the set of all sequences of 100 Californian voters would be the sample space Ω. We assume that sampling without replacement is used: only sequences of 100 *different* voters are allowed. For simplicity an ordered sample is considered, that is a sequence {Alice, Brian} is different from {Brian, Alice}. We also take for granted that each potential voter knows exactly his future choice, that is he/she doesn't choose randomly.

Alice knows only whether or not Arnold Schwarzenegger has received at least 60 votes. Her incomplete information is described by the σ-algebra $\mathcal{F}_{\text{Alice}}$ that contains: (1) the set of all sequences in Ω where at least 60 people vote for Schwarzenegger; (2) the set of all sequences where fewer than 60 vote for Schwarzenegger; (3) the whole sample space Ω; and (4) the empty set $\varnothing$.

Brian knows the exact number of voters who are going to vote for Schwarzenegger. His incomplete information is described by the corresponding partition $\Omega = B_0 \sqcup B_1 \dots \sqcup B_{100}$ (though some of these sets may be empty, depending on the Californian voters...) and the σ-algebra $\mathcal{F}_{\text{Brian}}$ consists of 2^{101} events.

In this case Alice's σ-algebra is a subset of Brian's: $\mathcal{F}_{\text{Alice}} \subset \mathcal{F}_{\text{Brian}}$. The Brian's σ-algebra is in turn the subset of the much larger "complete information" σ-algebra 2^Ω consisting of $2^{n(n-1)\dots(n-99)}$ events, where n is the number of all potential voters in California.

Non-Atomic Examples

Example 4

A number between 0 and 1 is chosen at random, uniformly. Here $\Omega = [0,1]$, is the σ-algebra of Borel sets on Ω, and P is the Lebesgue measure on $[0,1]$.

In this case the open intervals of the form (a,b), where $0 < a < b < 1$, could be taken as the generator sets. Each such set can be ascribed the probability of $P((a,b)) = (b - a)$, which generates the Lebesgue measure on $[0,1]$, and the Borel σ-algebra on Ω.

Example 5

A fair coin is tossed endlessly. Here one can take $\Omega = \{0,1\}^\infty$, the set of all infinite sequences of numbers 0 and 1. Cylinder sets $\{(x_1, x_2, ...) \in \Omega : x_1 = a_1, ..., x_n = a_n\}$ may be used as the generator sets. Each such set describes an event in which the first n tosses have resulted in a fixed sequence $(a_1, ..., a_n)$, and the rest of the sequence may be arbitrary. Each such event can be naturally given the probability of 2^{-n}.

These two non-atomic examples are closely related: a sequence $(x_1,x_2,...) \in \{0,1\}^\infty$ leads to the number $2^{-1}x_1 + 2^{-2}x_2 + ... \in [0,1]$. This is not a one-to-one correspondence between $\{0,1\}^\infty$ and $[0,1]$ however: it is an isomorphism modulo zero, which allows for treating the two probability spaces as two forms of the same probability space. In fact, all non-pathologic non-atomic probability spaces are the same in this sense. They are so-called standard probability spaces. Basic applications of probability spaces are insensitive to standardness. However, non-discrete conditioning is easy and natural on standard probability spaces, otherwise it becomes obscure.

Related Concepts

Probability Distribution

Any probability distribution defines a probability measure.

Random Variables

A random variable X is a measurable function $X: \Omega \rightarrow S$ from the sample space Ω to another measurable space S called the *state space*.

The notation $\Pr(X \in A)$ is a commonly used shorthand for $P(\{\Omega \in \Omega: X(\omega) \in A\})$.

Defining the Events in Terms of the Sample Space

If Ω is countable we almost always define $\mathcal{F}$ as the power set of Ω, i.e. $\mathcal{F} = 2^\Omega$ which is trivially a σ-algebra and the biggest one we can create using Ω. We can therefore omit 1! and just write (Ω,P) to define the probability space.

On the other hand, if Ω is uncountable and we use $\mathcal{F} = 2^\Omega$ we get into trouble defining our probability measure P because $\mathcal{F}$ is too "large", i.e. there will often be sets to which it will be impossible to assign a unique measure. In this case, we have to use a smaller σ-algebra $\mathcal{F}$, for example the Borel algebra of Ω, which is the smallest σ-algebra that makes all open sets measurable.

Conditional Probability

Kolmogorov's definition of probability spaces gives rise to the natural concept of conditional probability. Every set A with non-zero probability (that is, $P(A) > 0$) defines another probability measure

$$P(B \mid A) = \frac{P(B \cap A)}{P(A)}$$

On the space. This is usually pronounced as the "probability of B given A". For any event B such that $P(B) > 0$ the function Q defined by $Q(A) = P(A \mid B)$ for all events A is itself a probability measure.

Independence

Two events, A and B are said to be independent if $P(A) \cap B)=P(A)P(B)$.

Two random variables, X and Y, are said to be independent if any event defined in terms of X is independent of any event defined in terms of Y. Formally, they generate independent σ-algebras, where two σ-algebras G and H, which are subsets of F are said to be independent if any element of G is independent of any element of H.

Mutual Exclusivity

Two events, A and B are said to be mutually exclusive or *disjoint* if $P(A \cap B) = 0$. (This is weaker than $A \cap B = \varnothing$, which is the definition of disjoint for sets).

If A and B are disjoint events, then $P(A \cup B) = P(A) + P(B)$. This extends to a (finite or countably infinite) sequence of events. However, the probability of the union of an uncountable set of events is not the sum of their probabilities. For example, if Z is a normally distributed random variable, then $P(Z=x)$ is 0 for any x, but $P(Z \in \mathrm{R}) = 1$.

The event $A \cap B$ is referred to as "A and B", and the event $A \cup B$ as "A or B".

Event (Probability Theory)

In probability theory, an event is a set of outcomes of an experiment (a subset of the sample space) to which a probability is assigned. A single outcome may be an element of many different events, and different events in an experiment are usually not equally likely, since they may include very different groups of outcomes. An event defines a complementary event, namely the complementary set (the event *not* occurring), and together these define a Bernoulli trial: did the event occur or not?

Typically, when the sample space is finite, any subset of the sample space is an event (*i.e.* all elements of the power set of the sample space are defined as events). However, this approach does not work well in cases where the sample space is uncountably infinite, most notably when the outcome is a real number. So, when defining a probability space it is possible, and often necessary, to exclude certain subsets of the sample space from being events.

A Simple Example

If we assemble a deck of 52 playing cards with no jokers, and draw a single card from the deck, then the sample space is a 52-element set, as each card is a possible outcome. An event, however, is any subset of the sample space, including any singleton set (an elementary event), the empty set (an impossible event, with probability zero) and the sample space itself (a certain event, with probability one). Other events are proper subsets of the sample space that contain multiple elements. So, for example, potential events include:

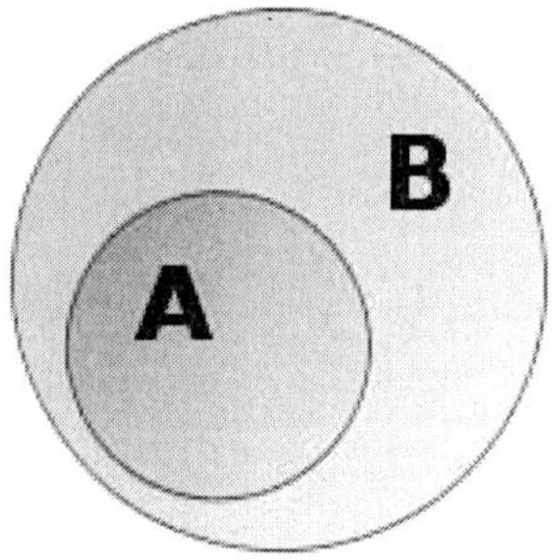

Figure: *A Venn diagram of an event.* B *is the sample space and* A *is an event.*

By the ratio of their areas, the probability of *A* is approximately 0.4.

- "Red and black at the same time without being a joker" (0 elements),
- "The 5 of Hearts" (1 element),
- "A King" (4 elements),
- "A Face card" (12 elements),
- "A Spade" (13 elements),
- "A Face card or a red suit" (32 elements),
- "A card" (52 elements).

Since all events are sets, they are usually written as sets (e.g. {1, 2, 3}), and represented graphically using Venn diagrams. Given that each outcome in the sample space Ω is equally likely, the probability of an event A is the following formula:

$$\mathrm{P}(A)=\frac{|A|}{|\Omega|}\left(\text{alternatively: }\Pr(A)=\frac{|A|}{|\Omega|}\right)$$

This rule can readily be applied to each of the example events above.

Events in Probability Spaces

Defining all subsets of the sample space as events works well when there are only finitely many outcomes, but gives rise to problems when the sample space is infinite. For many standard probability distributions, such as the normal distribution, the sample space is the set of real numbers or some subset of the real numbers. Attempts to define probabilities for all subsets of the real numbers run into difficulties when one considers 'badly behaved' sets, such as those that are nonmeasurable. Hence, it is necessary to restrict attention to a more limited family of subsets. For the standard tools of probability theory, such as joint and conditional probabilities, to work, it is necessary to use a σ-algebra, that is, a family closed under complementation and countable unions of its members. The most natural choice is the Borel measurable set derived from unions and intersections of intervals. However, the larger class of Lebesgue measurable sets proves more useful in practice.

In the general measure-theoretic description of probability spaces, an event may be defined as an element of a selected σ-algebra of subsets of the sample space. Under this definition, any subset of the sample space that is not an element of the σ-algebra is not an event, and does not have a probability. With a reasonable specification of the probability space, however, all *events of interest* are elements of the σ-algebra.

A Note on Notation

Even though events are subsets of some sample space Ω, they are often written as propositional formulas involving random variables. For example, if X is a real-valued random variable defined on the sample space Ω, the event

$$\{\omega\in \Omega \mid u < X(\omega) \leq v\}$$

can be written more conveniently as, simply,

$$u < X \leq v.$$

This is especially common in formulas for a probability, such as

$$\Pr(u < X \leq v) = F(v) - F(u).$$

The set $u < X \leq v$ is an example of an inverse image under the mapping X because $\omega \in X^{-1}((u, v])$ if and only if $u < X(\omega) \leq v$.

Sample Space

In probability theory, the sample space of an experiment or random trial is the set of all possible outcomes or results of that experiment. A sample space is usually denoted using set notation, and the possible outcomes are listed as elements in the set. It is common to refer to a sample space by the labels S, Ω, or U (for "universal set").

For example, if the experiment is tossing a coin, the sample space is typically the set {head, tail}. For tossing two coins, the corresponding sample space would be {(head,head), (head,tail), (tail,head), (tail,tail)}. For tossing a single six-sided die, the typical sample space is {1, 2, 3, 4, 5, 6} (in which the result of interest is the number of pips facing up).

A well-defined sample space is one of three basic elements in a probabilistic model (a probability space); the other two are a well-defined set of possible events (a sigma-algebra) and a probability assigned to each event (a probability measure function).

Multiple Sample Spaces

For many experiments, there may be more than one plausible sample space available, depending on what result is of interest to the experimenter. For example, when drawing a card from a standard deck of fifty-two playing cards, one possibility for the sample space could be the various ranks (Ace through King), while another could be the suits (clubs, diamonds, hearts, or spades). A more complete description of outcomes, however, could specify both the denomination and the suit, and a sample space describing each individual card can be constructed as the Cartesian product of the two sample spaces noted above (this space would contain fifty-two equally likely outcomes). Still other sample spaces are possible, such as {right-side up, up-side down} if some cards have been flipped when shuffling.

Equally Likely Outcomes

In some sample spaces, it is reasonable to estimate or assume that all outcomes in the space are equally likely (that they occur with equal probability). For example, when tossing an ordinary coin, one typically assumes that the outcomes "head" and "tail" are equally likely to occur. An implicit assumption that all outcomes in the sample space are equally likely underpins most randomization tools used in common games of chance (e.g. rolling dice, shuffling cards, spinning tops or wheels, drawing lots, etc.). Of course, players in such games can try to cheat by subtly introducing systematic deviations from equal likelihood (e.g. with marked cards, loaded or shaved dice, and other methods).

Some treatments of probability assume that the various outcomes of an experiment are always defined so as to be equally likely. However, there are experiments that are not easily described by a sample space of equally likely outcomes— for example, if one were to toss a thumb tack many times and observe whether it landed with its point upward or downward, there is no symmetry to suggest that the two outcomes should be equally likely.

Though most random phenomena do not have equally likely outcomes, it can be helpful to define a sample space in such a way that outcomes are at least approximately equally likely, since this condition significantly simplifies the computation of probabilities for events within the sample space. If each individual outcome occurs with the same probability, then the probability of any event becomes simply:

$$P(event) = \frac{\text{number of outcomes in event}}{\text{number of outcomes in sample space}}$$

Simple Random Sample

In statistics, inferences are made about characteristics of a population by studying a sample of that population's individuals. In order to arrive at a sample that presents an unbiased estimate of the true characteristics of the population, statisticians often seek to study a simple random sample— that is, a sample in which every individual in the population is equally likely to be included. The result of this is that every possible combination of individuals who could be chosen for the sample is also equally likely (that is, the space of simple random samples of a given size from a given population is composed of equally likely outcomes).

Infinitely Large Sample Spaces

In an elementary approach to probability, any subset of the sample space is usually called an event. However, this gives rise to problems when the sample space is infinite, so that a more precise definition of an event is necessary. Under this definition only measurable subsets of the sample space, constituting a σ-algebra over the sample space itself, are considered events. However, this has essentially only theoretical significance, since in general the σ-algebra can always be defined to include all subsets of interest in applications.

Elementary Event

In probability theory, an elementary event (also called an atomic event or simple event) is an event which contains only a single outcome in the sample space. Using set theory terminology, an elementary event is a singleton. Elementary events and their corresponding outcomes are often written interchangeably for simplicity, as such an event corresponds to precisely one outcome.

The following are examples of elementary events:

- All sets $\{k\}$, where $k \in$ N if objects are being counted and the sample space is $S = \{0, 1, 2, 3, ...\}$ (the natural numbers).
- {HH}, {HT}, {TH} and {TT} if a coin is tossed twice. S = {HH, HT, TH, TT}. H stands for heads and T for tails.
- All sets $\{x\}$, where x is a real number. Here X is a random variable with a normal distribution and $S = (-\infty, +\infty)$. This example shows that, because the probability of each elementary event is zero, the probabilities assigned to elementary events do not determine a continuous probability distribution.

Probability of an Elementary Event

Elementary events may occur with probabilities that are between zero and one (inclusively). In a discrete probability distribution whose sample space is finite, each elementary event is assigned a particular probability. In contrast, in a continuous distribution, individual elementary events must all have a probability of zero because there are infinitely many of them— then non-zero probabilities can only be assigned to non-elementary events.

Some "mixed" distributions contain both stretches of continuous elementary events and some discrete elementary events; the discrete elementary events in such distributions can be called atoms or atomic events and can have non-zero probabilities.

Under the measure-theoretic definition of a probability space, the probability of an elementary event need not even be defined. In particular, the set of events on which probability is defined may be some σ-algebra on S and not necessarily the full power set.

Random Variable

In probability and statistics, a random variable, aleatory variable or stochastic variable is a variable whose value is subject to variations due to chance (i.e. randomness, in a mathematical sense). A random variable can take on a set of possible different values (similarly to other mathematical variables), each with an associated probability, in contrast to other mathematical variables.

A random variable's possible values might represent the possible outcomes of a yet-to-be-performed experiment, or the possible outcomes of a past experiment whose already-existing value is uncertain (for example, as a result of incomplete information or imprecise measurements). They may also conceptually represent either the results of an "objectively" random process (such as rolling a die) or the "subjective" randomness that results from incomplete knowledge of a quantity. The meaning of the probabilities assigned to the potential values of a random variable is not part of probability theory itself but is instead related to philosophical arguments over the interpretation of probability. The mathematics works the same regardless of the particular interpretation in use.

The mathematical function describing the possible values of a random variable and their associated probabilities is known as a probability distribution. *Random variables* can be *discrete*, that is, taking any of a specified finite or countable list of values, endowed with a probability mass function, characteristic of a probability distribution; or *continuous*, taking any numerical value in an interval or collection of intervals, via a probability density function that is characteristic of a probability distribution; or a mixture of both types. The realisations of a random variable, that is, the results of randomly choosing values according to the variable's probability distribution function, are called random variates.

The formal mathematical treatment of random variables is a topic in probability theory. In that context, a random variable is understood as a function defined on a sample space whose outputs are numerical values.

Definition

A *random variable* $X : \Omega \to E$ is a measurable function from the set of possible outcomes Ω to some set E. Usually, $E = \mathbb{R}$, otherwise the term *random element* is used instead. The technical axiomatic definition requires both Ω and E to be measurable spaces.

As a real-valued function, X often describes some numerical quantity of a given event. E.g. the number of heads after a certain number of coin flips; the heights of different people.

When the image (or range) of X is finite or countably infinite, the random variable is called a discrete random variable and its distribution can be described by a probability mass function which assigns a probability to each value in the image of X. If the image is uncountably infinite then X is called a continuous random variable. In the special case that it is absolutely continuous, its distribution can be described by a probability density function, which assigns probabilities to intervals; in particular, each individual point must necessarily have probability zero for an absolutely continuous random variable. Not all continuous random variables are absolutely continuous, for example a mixture distribution. Such random variables cannot be described by a probability density or a probability mass function.

All random variables can be described by their cumulative distribution function, which describes the probability that the random variable will be less than or equal to a certain value.

Examples

For example, in an experiment a person may be chosen at random, and one random variable may be the person's height. Mathematically, the random variable is interpreted as a function which maps the person to the person's height. Associated with the random variable is a probability distribution that allows the computation of the probability that the height is in any non-pathological subset of possible values, such as probability that the height is between 180 and 190 cm, or the probability that the height is either less than 150 or more than 200 cm.

Another random variable may be the person's number of children; this is a discrete random variable with non-negative integer values. It allows the computation of probabilities for individual integer values – the probability mass function (PMF) – or for sets of values, including infinite sets. For example, the event of interest may be "an even number of children". For both finite and infinite event sets, their probabilities can be found by adding up the PMFs of the elements; that is, the probability of an even number of children is the infinite sum PMF(0) + PMF(2) + PMF(4) + ...

In examples such as these, the sample space (the set of all possible persons) is often suppressed, since it is mathematically hard to describe, and the possible values of the random variables are then treated as a sample space. But when two random variables are measured on the same sample space of outcomes, such as the height and number of children being computed on the same random persons, it is easier to track their relationship if it is acknowledged that both height and number of children come from the same random person, for example so that questions of whether such random variables are correlated or not can be posed.

Extensions

The basic concept of "random variable" in statistics is real-valued, and therefore expected values, variances and other measures can be computed. However, one can consider arbitrary types such as boolean values, categorical variables, complex numbers, vectors, matrices, sequences, trees, sets, shapes, manifolds, functions, and processes. The term *random element* is used to encompass all such related concepts.

Another extension is the stochastic process, a set of indexed random variables (typically indexed by time or space).

These more general concepts are particularly useful in fields such as computer science and natural language processing where many of the basic elements of analysis are non-numerical. Such general random elements can sometimes be treated as sets of real-valued random variables — often more specifically as random vectors. For example:

- A "random word" may be parameterized by an integer-valued index into the vocabulary of possible words; alternatively, as an indicator vector, in which exactly one element is a 1, and the others are 0, with the one indexing a particular word into a vocabulary.
- A "random sentence" may be parameterized as a vector of random words.
- A random graph, for a graph with V edges, may be parameterized as an *NxN* matrix, indicating the weight for each edge, or 0 for no edge. (If the graph has no weights, 1 indicates an edge; 0 indicates no edge.)

Reduction to numerical values is not essential for dealing with random elements: a randomly selected individual remains an individual, not a number.

Examples

The possible outcomes for one coin toss can be described by the sample space $\Omega = \{\text{heads}, \text{tails}\}$. We can introduce a real-valued random variable Y that models a $1 payoff for a successful bet on heads as follows:

$$Y(\omega) = \begin{cases} 1, & \text{if } \omega = \text{heads}, \\ \\ 0, & \text{if } \omega = \text{tails}. \end{cases}$$

If the coin is equally likely to land on either side then Y has a probability mass function f_Y given by:

$$f_Y(y) = \begin{cases} \frac{1}{2}, & \text{if } y = 1, \\ \\ \frac{1}{2}, & \text{if } y = 0, \end{cases}$$

A random variable can also be used to describe the process of rolling dice and the possible outcomes. The most obvious representation for the two-dice case is to take the set of pairs of numbers n_1 and n_2 from {1, 2, 3, 4, 5, 6} representing the numbers on the two dice as the sample space, defining the random variable X to be equal to the total number rolled, the sum of the numbers in each pair.

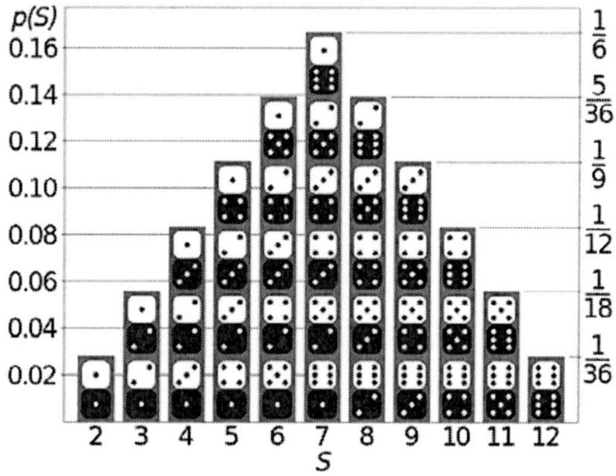

Figure: *If the sample space is the set of possible numbers rolled on two dice, and the random variable of interest is the sum* S *of the numbers on the two dice, then* S *is a discrete random variable whose distribution is described by the probability mass function plotted as the height of picture columns here.*

In this case, the random variable of interest X is defined as the function that maps the pair to the sum:

$$X((n_1, n_2)) = n_1 + n_2$$

and has probability mass function f_X given by:

$$f_X(S) = \frac{\min(S-1, 13-S)}{36}, \text{for } S \in \{2,3,4,5,6,7,8,9,10,11,12\}$$

An example of a continuous random variable would be one based on a spinner that can choose a horizontal direction. Then the values taken by the random variable are directions. We could represent these directions by North, West, East, South, Southeast, etc. However, it is commonly more convenient to map the sample space to a random variable which takes values which are real numbers. This can be done, for example, by mapping a direction to a bearing in degrees clockwise from North. The random variable then takes values which are real numbers from the interval [0, 360), with all parts of the range being "equally likely". In this case, X = the angle spun. Any real number has probability zero of being selected, but a positive probability can be assigned to any *range* of values. For example, the probability of choosing a number in [0, 180] is $^1/_2$. Instead of speaking of a probability mass function, we say that the probability *density* of X is 1/360. The probability of a subset of [0, 360) can be calculated by multiplying the measure of the set by 1/360. In general, the probability of a set for a given continuous random variable can be calculated by integrating the density over the given set.

An example of a random variable of mixed type would be based on an experiment where a coin is flipped and the spinner is spun only if the result of the coin toss is heads. If the result is tails, $X = -1$; otherwise X = the value of the spinner as in the preceding example. There is a probability of $^1/_2$ that this random variable will have the value –1. Other ranges of values would have half the probability of the last example.

Measure-Theoretic Definition

The most formal, axiomatic definition of a random variable involves measure theory. Continuous random variables are defined in terms of sets of numbers, along with functions that map such sets to probabilities. Because of various difficulties (e.g. the Banach–Tarski paradox) that arise if such sets are insufficiently constrained, it is necessary to introduce what is termed a sigma-algebra to constrain the possible sets over which probabilities can be defined. Normally, a particular such sigma-algebra is used, the Borel σ-algebra, which allows for probabilities to be defined over any sets that can be derived either directly from continuous intervals of numbers or by a finite or countably infinite number of unions and/or intersections of such intervals.

The Measure-Theoretic Definition is as Follows

Let $(\Omega, \mathcal{F}, P)$ be a probability space and $(E, \mathcal{E})$ a measurable space. Then an $(E, \mathcal{E})$-valued random variable is a function $X : \Omega \to E$ which is $(\mathcal{F}, \mathcal{E})$-measurable. The latter means that, for every subset $B \in \mathcal{E}$, its preimage $X^{-1}(B) \in \mathcal{F}$ where $X^{-1}(B) = \{\omega : X(\omega) \in B\}$. This definition enables us to measure any subset $B \in \mathcal{E}$ in the target space by looking at its preimage, which by assumption is measurable.

When E is a topological space, then the most common choice for the σ-algebra $\mathcal{E}$ is the Borel σ-algebra $\mathcal{B}(E)$, which is the σ-algebra generated by the collection of all open sets in E. In such case the $(E, \mathcal{E})$-valued random variable is called the E-valued random variable.

Moreover, when space E is the real line $\mathbb{R}$, then such real-valued random variable is called simply the random variable.

Real-Valued Random Variables

In this case the observation space is the real numbers. Recall, $(\Omega, \mathcal{F}, P)$ is the probability space. For real observation space, the function $X : \Omega \rightarrow \mathbb{R}$ is a real-valued random variable if

$$\{\omega : X(\omega) \leq r\} \in \mathcal{F} \qquad \forall r \in \mathbb{R}.$$

This definition is a special case of the above because the set $\{(-\infty, r] : r \in \mathbb{R}\}$ generates the Borel σ-algebra on the real numbers, and it suffices to check measurability on any generating set. Here we can prove measurability on this generating set by using the fact that

Distribution Functions of Random Variables

If a random variable $X : \Omega \rightarrow \mathbb{R}$ defined on the probability space $(\Omega, \mathcal{F}, P)$ is given, we can ask questions like "How likely is it that the value of X is equal to 2?". This is the same as the probability of the event $\{\omega : X(\omega) = 2\}$ which is often written as $P(X = 2)$ or $p_X(2)$ for short.

Recording all these probabilities of output ranges of a real-valued random X variable yields the probability distribution of *X*. The probability distribution "forgets" about the particular probability space used to define X and only records the probabilities of various values of X. Such a probability distribution can always be captured by its cumulative distribution function

$$F_X(x) = \mathrm{P}(X \leq x)$$

and sometimes also using a probability density function, p_X . In measure-theoretic terms, we use the random variable X to "push-forward" the measure P on Ω to a measure p_X on $\mathbb{R}$. The underlying probability space Ω is a technical device used to guarantee the existence of random variables, sometimes to construct them, and to define notions such as correlation and dependence or independence based on a joint distribution of two or more random variables on the same probability space. In practice, one often disposes of the space Ω altogether and just puts a measure on $\mathbb{R}$ that assigns measure 1 to the whole real line, i.e., one works with probability distributions instead of random variables.

Moments

The probability distribution of a random variable is often characterised by a small number of parameters, which also have a practical interpretation. For example, it is often enough to know what its "average value" is. This is captured by the mathematical concept of expected value of a random variable, denoted E[X], and also called the first moment. In general, E[$f(X)$] is not equal to f(E[X]). Once the "average value" is known, one could then ask how far from this average value the values of X typically are, a question that is answered by the variance and standard deviation of a random variable. E[X] can be viewed intuitively as an average obtained from an infinite population, the members of which are particular evaluations of X.

Mathematically, this is known as the (generalised) problem of moments: for a given class of random variables X, find a collection $\{f_i\}$ of functions such that the expectation values E[$f_i(X)$] fully characterise the distribution of the random variable X. Moments can only be defined for real-valued functions of random variables (or complex-valued, etc.). If the random variable is itself real-valued, then moments of the variable itself can be taken, which are equivalent to moments of the identity function $f(X) = X$ of the random variable. However, even for non-real-valued random variables, moments can be taken of real-valued functions of those variables. For example, for a categorical random variable X that can take on the nominal values "red", "blue" or "green", the real-valued function $[X = \text{green}]$ can be constructed; this uses the Iverson bracket, and has the value 1 if X has the value "green", 0 otherwise. Then, the expected value and other moments of this function can be determined.

Functions of Random Variables

A new random variable Y can be defined by applying a real Borel measurable function $g : \mathbb{R} \to \mathbb{R}$ to the outcomes of a real-valued random variable X. The cumulative distribution function of Y is

$$F_Y(y) = \mathrm{P}(g(X) \le y).$$

If function g is invertible, i.e. g^{-1} exists, and is either increasing or decreasing, then the previous relation can be extended to obtain

$$F_Y(y) = \mathrm{P}(g(X) \le y) = \begin{cases} \mathrm{P}(X \le g^{-1}(y)) = F_X(g^{-1}(y)), & \text{if } g^{-1} \text{ increasing,} \\ \\ \mathrm{P}(X \ge g^{-1}(y)) = 1 - F_X(g^{-1}(y)), & \text{if } g^{-1} \text{ decreasing.} \end{cases}$$

and, again with the same hypotheses of invertibility of g, assuming also differentiability, we can find the relation between the probability density functions by differentiating both sides with respect to y, in order to obtain

$$f_Y(y) = f_X(g^{-1}(y)) \left| \frac{dg^{-1}(y)}{dy} \right|.$$

If there is no invertibility of g but each y admits at most a countable number of roots (i.e. a finite, or countably infinite, number of x_i such that $y = g(x_i)$) then the previous relation between the probability density functions can be generalized with

$$f_Y(y) = \sum_i f_X(g_i^{-1}(y)) \left| \frac{dg_i^{-1}(y)}{dy} \right|$$

where $x_i = g_i^{-1}(y)$. The formulas for densities do not demand g to be increasing. In the measure-theoretic, axiomatic approach to probability, if we have a random variable X on Ω and a Borel measurable function $g : \mathbb{R} \rightarrow \mathbb{R}$, $Y = g(X)$ then will also be a random variable on Ω , since the composition of measurable functions g is also measurable. (However, this is not true if is Lebesgue measurable.) The same procedure that allowed one (Ω, P) to $(\mathbb{R}, dF_X)$ go from a probability space to can be used to obtain the distribution of .

Example 1

Let X be a real-valued, continuous random variable and let $Y = X^2$.

$$F_Y(y) = \mathrm{P}(X^2 \leq y).$$

If $y < 0$, then P(X^2 dd y) = 0, so

$$F_Y(y) = 0 \qquad \text{if} \quad y < 0.$$

If ≥ ed 0, then

$$\mathrm{P}(X^2 \leq y) = \mathrm{P}(|X| \leq \sqrt{y}) = \mathrm{P}(-\sqrt{y} \leq X \leq \sqrt{y}),$$

so

$$F_Y(y) = F_X(\sqrt{y}) - F_X(-\sqrt{y}) \qquad \text{if} \quad y \geq 0.$$

Example 2

Suppose X is a random variable with a cumulative distribution

$$F_X(x) = P(X \leq x) = \frac{1}{(1 + e^{-x})^{\theta}}$$

where $\theta > 0$ is a fixed parameter. Consider the random variable $Y = \log(1 + e^{-X})$. Then,

$$F_Y(y) = P(Y \leq y) = P(\log(1 + e^{-X}) \leq y) = P(X > -\log(e^y - 1)).$$

The last expression can be calculated in terms of the cumulative distribution of X,so

$$F_Y(y) = 1 - F_X(-\log(e^y - 1))$$

$$= 1 - \frac{1}{(1 + e^{\log(e^y - 1)})^{\theta}}$$

$$= 1 - \frac{1}{(1 + e^y - 1)^{\theta}} = 1 - e^{-y\theta}.$$

Example 3

Suppose X is a random variable with a standard normal distribution, whose density is

$$f_X(x) = \frac{1}{\sqrt{2\pi}} e^{-x^2/2}.$$

Consider the random variable $Y = X^2$. We can find the density using the above formula for a change of variables:

$$f_Y(y) = \sum_i f_X(g_i^{-1}(y)) \left| \frac{dg_i^{-1}(y)}{dy} \right|.$$

In this case the change is not monotonic, because every value of Y has two corresponding values of X (one positive and negative). However, because of symmetry, both halves will transform identically, i.e.

$$f_Y(y) = 2 f_X(g^{-1}(y)) \left| \frac{dg^{-1}(y)}{dy} \right|.$$

The inverse transformation is

$$x = g^{-1}(y) = \sqrt{y}$$

and its derivative is

$$\frac{dg^{-1}(y)}{dy} = \frac{1}{2\sqrt{y}}.$$

Then:

$$f_Y(y) = 2 \frac{1}{\sqrt{2\pi}} e^{-y/2} \frac{1}{2\sqrt{y}} = \frac{1}{\sqrt{2\pi y}} e^{-y/2}.$$

This is a chi-squared distribution with one degree of freedom.

Equivalence of Random Variables

There are several different senses in which random variables can be considered to be equivalent. Two random variables can be equal, equal almost surely, or equal in distribution. In increasing order of strength, the precise definition of these notions of equivalence is given below.

Equality in Distribution

If the sample space is a subset of the real line, random variables X and Y are *equal in distribution* (denoted $X \stackrel{d}{=} Y$) if they have the same distribution functions:

$$P(X \leq x) = P(Y \leq x) \quad \text{for all} \quad x.$$

Two random variables having equal moment generating functions have the same distribution. This provides, for example, a useful method of checking equality of certain functions of i.i.d. random variables. However, the moment generating function exists only for distributions that have a defined Laplace transform.

Almost Sure Equality

Two random variables X and Y are *equal almost surely* if, and only if, the probability that they are different is zero:

$$P(X \neq Y) = 0.$$

For all practical purposes in probability theory, this notion of equivalence is as strong as actual equality. It is associated to the following distance:

$$d_{\infty}(X,Y) = \operatorname*{ess\,sup}_{\omega} | X(\omega) - Y(\omega) |,$$

where "ess sup" represents the essential supremum in the sense of measure theory.

Equality

Finally, the two random variables X and Y are *equal* if they are equal as functions on their measurable space:

$$X(\omega) = Y(\omega) \qquad \text{for all } \omega.$$

Convergence

A significant theme in mathematical statistics consists of obtaining convergence results for certain sequences of random variables; for instance the law of large numbers and the central limit theorem.

There are various senses in which a sequence (X_n) of random variables can converge to a random variable X. These are explained in the article on convergence of random variables.

Chapter 5

Probability Measure

In mathematics, a probability measure is a real-valued function defined on a set of events in a probability space that satisfies measure properties such as *countable additivity*. The difference between a probability measure and the more general notion of measure (which includes concepts like area or volume) is that a probability measure must assign value 1 to the entire probability space.

Intuitively, the additivity property says that the probability assigned to the union of two disjoint events by the measure should be the sum of the probabilities of the events, e.g. the value assigned to "1 or 2" in a throw of a die should be the sum of the values assigned to "1" and "2".

Probability measures have applications in diverse fields, from physics to finance and biology.

Definition

The requirements for a function μ to be a probability measure on a probability space are that:

- μ must return results in the unit interval [0, 1], returning 0 for the empty set and 1 for the entire space.
- μ must satisfy the *countable additivity* property that for all countable collections $\{E_i\}$ of pairwise disjoint sets:

$$\mu\Big(\bigcup_{i\in I} E_i\Big) = \sum_{i\in I} \mu(E_i).$$

For example, given three elements 1, 2 and 3 with probabilities 1/4, 1/4 and 1/2, the value assigned to {1, 3} is 1/4 + 1/2 = 3/4, as in the diagram on the right. The conditional probability based on the intersection of events defined as:

$$P(B \mid A) = \frac{P(A \cap B)}{P(A)}.$$

satisfies the probability measure requirements so long as $P(A)$ is not zero.

Probability measures are distinct from the more general notion of fuzzy measures in which there is no requirement that the fuzzy values sum up to 1, and the additive property is replaced by an order relation based on set inclusion.

Example Applications

Market measures which assign probabilities to financial market spaces based on actual market movements are examples of probability measures which are of interest in mathematical finance, e.g. in the pricing of financial derivatives. For instance, a risk-neutral measure is a probability measure which assumes that the current value of assets is the expected value of the future payoff taken with respect to that same risk neutral measure (i.e. calculated using the corresponding risk neutral density function), and discounted at the risk-free rate. If there is a unique probability measure that must be used to price assets in a market, then the market is called a complete market.

Not all measures that intuitively represent chance or likelihood are probability measures. For instance, although the fundamental concept of a system in statistical mechanics is a measure space, such measures are not always probability measures. In general, in statistical physics, if we consider sentences of the form "the probability of a system S assuming state A is p" the geometry of the system does not always lead to the definition of a probability measure under congruence, although it may do so in the case of systems with just one degree of freedom.

Probability measures are also used in mathematical biology. For instance, in comparative sequence analysis a probability measure may be defined for the likelihood that a variant may be permissible for an amino acid in a sequence.

Bayes' Theorem

In probability theory and statistics, Bayes' theorem (alternatively Bayes' law or Bayes' rule) relates current to prior belief. It also relates current to prior evidence. It is important in the mathematical manipulation of conditional probabilities. Bayes' rule can be derived from more basic axioms of probability, specifically conditional probability.

When applied, the probabilities involved in Bayes' theorem may have any of a number of probability interpretations. In one of these interpretations, the theorem is used directly as part of a particular approach to statistical inference. ln particular, with the Bayesian interpretation of probability, the theorem expresses how a subjective degree of belief should rationally change to account for evidence: this is Bayesian inference, which is fundamental to Bayesian statistics. However, Bayes's theorem has applications in a wide range of calculations involving probabilities, not just in Bayesian inference.

Bayes's theorem is named after Rev. Thomas Bayes (/Èbejz/; 1701–1761), who first showed how to use new evidence to update beliefs. Bayes' unpublished manuscript was significantly edited by Richard Price before it was posthumously read at the Royal Society. Bayes' algorithm remained unknown until it was independently rediscovered and further developed by Pierre-Simon Laplace, who first published the modern formulation in his 1812 *Théorie analytique des probabilités.*

Sir Harold Jeffreys put Bayes' algorithm and Laplace's formulation on an axiomatic basis. Jeffreys wrote that Bayes's theorem "is to the theory of probability what Pythagoras's theorem is to geometry".

Introduction

Bayes's theorem is stated mathematically as the following simple form:

$$P(A \mid B) = \frac{P(B \mid A)P(A)}{P(B)}.$$

For proposition A and evidence or background B,

- $P(A)$, the prior probability, is the initial degree of belief in A.
- $P(A \mid B)$, the conditional probability, is the degree of belief in A, having taken B into account.
- the quotient $P(B \mid A)/P(B)$ represents the support B provides for A.

Another form of Bayes's Theorem that is generally encountered when looking at two competing statements or hypotheses is:

$$P(A \mid B) = \frac{P(B \mid A)P(A)}{P(B \mid A)P(A) + P(B \mid \neg A)P(\neg A)}.$$

For an epistemological interpretation:

For proposition A and evidence or background B,

- $P(A)$,the prior probability, is the initial degree of belief in A.
- $P(-A)$, is the corresponding probability of the initial degree of belief against A: $1-P(A)=P(-A)$
- $P(B \mid A)$, the conditional probability or likelihood, is the degree of belief in B, given that the proposition A is true.
- $P(B \mid -A)$, the conditional probability or likelihood, is the degree of belief in B, given that the proposition A is false.
- $P(A \mid B)$, the posterior probability, is the probability for A after taking into account B for and against A.

Introductory Example (Epistemological Interpretation)

Suppose a man told you he had a nice conversation with someone on the train. Not knowing anything about this conversation, the prior probability that he was speaking to a woman is 50% (assuming the speaker was as likely to strike up a conversation with a man as with a woman). Now suppose he also told you that his conversational partner had long hair. It is now more likely he was speaking to a woman, since women are more likely to have long hair than men. Bayes's theorem can be used to calculate the probability that the person was a woman.

To see how this is done, let W represent the event that the conversation was held with a woman, and L denote the event that the conversation was held with a long-haired person. For this example it can be assumed that women constitute half the population. So, not knowing anything else, the prior probability that W occurs is $P(W) = 0.5$.

Suppose it is also known that 75% of women have long hair, which we denote as $P(L \mid W) = 0.75$ (read: the probability of event L given event W is 0.75, meaning that the probability of a person having long hair (event "L"), given that we already know that the person is a woman ("event W") is 75%). This is the conditional probability that it was a woman who had the conversation on the train. Likewise, suppose it is known that 15% of men have long hair, or $P(L \mid M) = 0.15$, where M is the complementary event of W, i.e., the event that the conversation was held with a man (assuming that every human is either a man or a woman). This is the conditional probability against its having been a woman who had the conversation on the train.

Our goal is to calculate the posterior probability that the conversation was held with a woman, given the fact that the person had long hair, or, in our notation, $P(W \mid L)$. Using the formula for Bayes's theorem, we have:

$$P(W \mid L) = \frac{P(L \mid W)P(W)}{P(L)} = \frac{P(L \mid W)P(W)}{P(L \mid W)P(W) + P(L \mid M)P(M)}$$

where we have used the law of total probability to expand $P(L)$. The numeric answer can be obtained by substituting the above values into this formula (the algebraic multiplication is annotated using " · ", the centred dot). This yields

$$P(W \mid L) = \frac{0.75 \cdot 0.50}{0.75 \cdot 0.50 + 0.15 \cdot 0.50} = \frac{5}{6} \approx 0.83,$$

i.e., the probability that the conversation was held with a woman, given that the person had long hair, is about 83%. More examples are provided below.

Another way to do this calculation is as follows. Initially, it is equally likely that the conversation is held with a woman as with a man, so the prior odds are 1:1. The respective chances that a man and a woman have long hair are 15% and 75%. It is 5 times more likely that a woman has long hair than that a man has long hair. We say that the likelihood ratio or Bayes factor is 5:1. Bayes's theorem in odds form, also known as Bayes's rule, tells us that the posterior odds that the person was a woman is also 5:1 (the prior odds, 1:1, times the likelihood ratio, 5:1). In a formula:

$$\frac{P(W \mid L)}{P(M \mid L)} = \frac{P(W)}{P(M)} \cdot \frac{P(L \mid W)}{P(L \mid M)}.$$

Statement and Interpretation

Mathematically, Bayes's theorem gives the relationship between the probabilities of A and B, $P(A)$ and $P(B)$, and the conditional probabilities of A given B and B given A, $P(A \mid B)$ and $P(B \mid A)$. In its most common form, it is:

$$P(A \mid B) = \frac{P(B \mid A)P(A)}{P(B)}.$$

The meaning of this statement depends on the interpretation of probability ascribed to the terms:

Bayesian Interpretation

In the Bayesian (or epistemological) interpretation, probability measures a *degree of belief*. Bayes's theorem then links the degree of belief in a proposition before and after accounting for evidence. For example, suppose it is believed with 50% certainty that a coin is twice as likely to land heads than tails. If the coin is flipped a number of times and the outcomes observed, that degree of belief may rise, fall or remain the same depending on the results.

Relative size	Case B	Case $\bar{B}$	Total
Condition A	w	x	$w+x$
Condition $\bar{A}$	y	z	$y+z$
Total	$w+y$	$x+z$	$w+x+y+z$

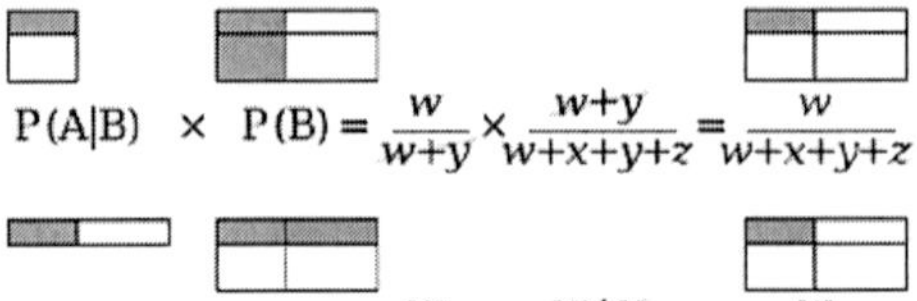

$$P(A|B) \times P(B) = \frac{w}{w+y} \times \frac{w+y}{w+x+y+z} = \frac{w}{w+x+y+z}$$

$$P(B|A) \times P(A) = \frac{w}{w+x} \times \frac{w+x}{w+x+y+z} = \frac{w}{w+x+y+z}$$

Figure: *A geometric visualisation of Bayes's theorem. In the table, the values* ax, ay, bx *and* by *give the relative weights of each corresponding condition and case. The figures denote the cells of the table involved in each metric, the probability being the fraction of each figure that is shaded. This shows that P(A | X) P(X) = P(X | A) P(A) i.e. P(A | X) = P(X | A) P(A) / P(X). Similar reasoning can be used to show that P(B | X) = P(X | B) P(B) / P(X) etc.*

For proposition *A* and evidence *B*,

- *P*(*A*), the *prior*, is the initial degree of belief in *A*.
- *P*(*A* | *B*), the *posterior*, is the degree of belief having accounted for *B*.
- the quotient *P*(*B* | *A*)/*P*(*B*) represents the support *B* provides for *A*.

Frequentist Interpretation

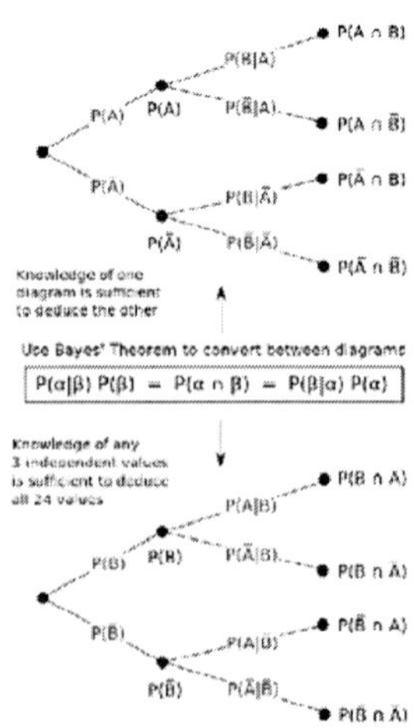

Figure: *Illustration of frequentist interpretation with tree diagrams. Bayes's theorem connects conditional probabilities to their inverses.*

In the frequentist interpretation, probability measures a *proportion of outcomes*. For example, suppose an experiment is performed many times. $P(A)$ is the proportion of outcomes with property A, and $P(B)$ that with property B. $P(B|A)$ is the proportion of outcomes with property B *out of* outcomes with property A, and $P(A|B)$ the proportion of those with A *out of* those with B.

The role of Bayes's theorem is best visualized with tree diagrams, as shown to the right. The two diagrams partition the same outcomes by A and B in opposite orders, to obtain the inverse probabilities. Bayes's theorem serves as the link between these different partitionings.

Forms

Events

Simple Form: For events A and B, provided that $P(B) \neq 0$,

$$P(A \mid B) = \frac{P(B \mid A)P(A)}{P(B)}.$$

In many applications, for instance in Bayesian inference, the event B is fixed in the discussion, and we wish to consider the impact of its having been observed on our belief in various possible events A. In such a situation the denominator of the last expression, the probability of the given evidence B, is fixed; what we want to vary is A. Bayes's theorem then shows that the posterior probabilities are proportional to the numerator:

$$P(A \mid B) \propto P(A) \cdot P(B \mid A) \text{ (proportionality over } A \text{ for given } B\text{).}$$

In words: posterior is proportional to prior times likelihood.

If events A_1, A_2, ..., are mutually exclusive and exhaustive, i.e., one of them is certain to occur but no two can occur together, and we know their probabilities up to proportionality, then we can determine the proportionality constant by using the fact that their probabilities must add up to one. For instance, for a given event A, the event A itself and its complement $\neg A$ are exclusive and exhaustive. Denoting the constant of proportionality by c we have

$$P(A \mid B) = c \cdot P(A) \cdot P(B \mid A) \text{ and } P(\neg A \mid B) = c \cdot P(\neg A) \cdot P(B \mid \neg A) \cdot$$

Adding these two formulas we deduce that

$$c = \frac{1}{P(A) \cdot P(B \mid A) + P(\neg A) \cdot P(B \mid \neg A)}.$$

Extended Form: Often, for some partition $\{A_j\}$ of the event space, the event space is given or conceptualized in terms of $P(A_j)$ and $P(B|A_j)$. It is then useful to compute $P(B)$ using the law of total probability:

$$P(B)=\sum_j P(B\mid A_j)P(A_j),$$

$$\Rightarrow P(A_i\mid B)=\frac{P(B\mid A_i)P(A_i)}{\sum_j P(B\mid A_j)P(A_j)}.$$

In the special case where A is a binary variable:

$$P(A\mid B)=\frac{P(B\mid A)P(A)}{P(B\mid A)P(A)+P(B\mid\neg A)P(\neg A)}.$$

Random Variables

Consider a sample space Ω generated by two random variables X and Y. In principle, Bayes's theorem applies to the events $A = \{X = x\}$ and $B = \{Y = y\}$. However, terms become 0 at points where either variable has finite probability density. To remain useful, Bayes's theorem may be formulated in terms of the relevant densities.

Simple Form

If X is continuous and Y is discrete,

$$f_X(x\mid Y=y)=\frac{P(Y=y\mid X=x)f_X(x)}{P(Y=y)}.$$

If X is discrete and Y is continuous,

$$P(X=x\mid Y=y)=\frac{f_Y(y\mid X=x)P(X=x)}{f_Y(y)}.$$

If both X and Y are continuous,

$$f_X(x\mid Y=y)=\frac{f_Y(y\mid X=x)f_X(x)}{f_Y(y)}.$$

Extended Form

A continuous event space is often conceptualized in terms of the numerator terms. It is then useful to eliminate the denominator using the law of total probability. For $f_Y(y)$, this becomes an integral:

$$f_Y(y)=\int_{-\infty}^{\infty} f_Y(y\mid X=\xi)f_X(\xi)d\xi.$$

Bayes's Rule

Bayes' rule is Bayes's theorem in odds form.

$$O(A_1:A_2\mid B)=O(A_1:A_2)\cdot\Lambda(A_1:A_2\mid B)$$

where

$$\Lambda(A_1 : A_2 \mid B) = \frac{P(B \mid A_1)}{P(B \mid A_2)}$$

is called the Bayes factor or likelihood ratio and the odds between two events is simply the ratio of the probabilities of the two events. Thus

$$O(A_1 : A_2) = \frac{P(A_1)}{P(A_2)},$$

$$O(A_1 : A_2 \mid B) = \frac{P(A_1 \mid B)}{P(A_2 \mid B)},$$

So the rule says that the posterior odds are the prior odds times the Bayes factor, or in other words, posterior is proportional to prior times likelihood.

Derivation

For Events: Bayes's theorem may be derived from the definition of conditional probability:

$$P(A \mid B) = \frac{P(A \cap B)}{P(B)}, \text{ if } P(B) \neq 0,$$

$$P(B \mid A) = \frac{P(A \cap B)}{P(A)}, \text{ if } P(A) \neq 0,$$

$$\Rightarrow P(A \cap B) = P(A \mid B)P(B) = P(B \mid A)P(A),$$

$$\Rightarrow P(A \mid B) = \frac{P(B \mid A)P(A)}{P(B)}, \text{ if } P(B) \neq 0.$$

For Random Variables

For two continuous random variables X and Y, Bayes's theorem may be analogously derived from the definition of conditional density:

$$f_X(x \mid Y = y) = \frac{f_{X,Y}(x, y)}{f_Y(y)}$$

$$f_Y(y \mid X = x) = \frac{f_{X,Y}(x, y)}{f_X(x)}$$

$$\Rightarrow f_X(x \mid Y = y) = \frac{f_Y(y \mid X = x) f_X(x)}{f_Y(y)}.$$

Examples

Frequentist Example:

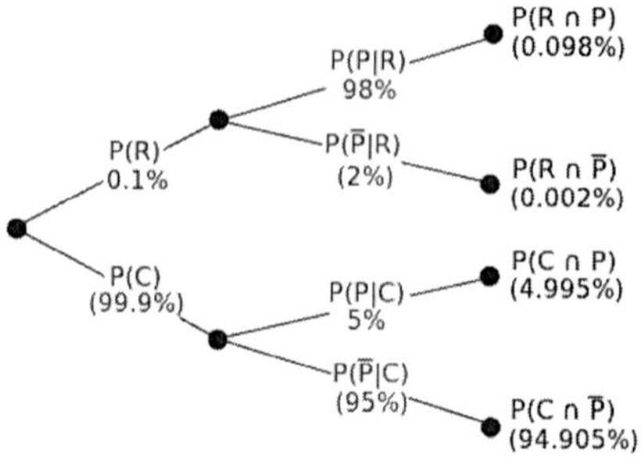

Figure: *Tree diagram illustrating frequentist example. R, C, P and P bar are the events representing rare, common, pattern and no pattern. Percentages in parentheses are calculated. Note that three independent values are given, so it is possible to calculate the inverse tree.*

An entomologist spots what might be a rare subspecies of beetle, due to the pattern on its back. In the rare subspecies, 98% have the pattern, or *P(Pattern | Rare)* = 98%. In the common subspecies, 5% have the pattern. The rare subspecies accounts for only 0.1% of the population. How likely is the beetle having the pattern to be rare, or what is *P(Rare | Pattern)*?

From the extended form of Bayes's theorem (since any beetle can be only rare or common),

$$\begin{aligned} P(\text{Rare} \mid \text{Pattern}) &= \frac{P(\text{Pattern} \mid \text{Rare})P(\text{Rare})}{P(\text{Pattern} \mid \text{Rare})P(\text{Rare}) + P(\text{Pattern} \mid \text{Common})P(\text{Common})} \\ &= \frac{0.98 \times 0.001}{0.98 \times 0.001 + 0.05 \times 0.999} \\ &\approx 1.9\%. \end{aligned}$$

Coin Flip Example

Concrete example from 5 August 2011 New York Times article by John Allen Paulos (quoted verbatim):

"Assume that you are presented with three coins, two of them fair and the other a counterfeit that always lands heads. If you randomly pick one of the three coins, the probability that it's the counterfeit is 1 in 3. This is the prior probability of the hypothesis that the coin is counterfeit. Now after picking the coin, you flip it three times and observe that it lands heads each time. Seeing this new evidence that your chosen coin has landed heads three times in a row, you want to know the revised posterior probability that it is the counterfeit. The answer to this question, found using Bayes's theorem (calculation mercifully omitted), is 4 in 5. You thus revise your probability estimate of the coin's being counterfeit upward from 1 in 3 to 4 in 5."

The calculation ("mercifully supplied") follows:

$$P(\text{Biased coin}) = \frac{1}{3}$$

$$P(\text{Fair coin}) = \frac{2}{3}$$

$$P(\text{H} \mid \text{Fair coin}) = \frac{1}{2}$$

$$P(\text{HHH} \mid \text{Fair coin}) = \frac{1}{2} \times \frac{1}{2} \times \frac{1}{2} = \frac{1}{8}$$

$$P(\text{HHH} \mid \text{Biased coin}) = 1 \times 1 \times 1 = 1$$

$$P(\text{Biased coin} \mid \text{HHH}) = \frac{P(\text{HHH} \mid \text{Biased coin})P(\text{Biased coin})}{P(\text{HHH} \mid \text{Biased coin})P(\text{Biased coin}) + P(\text{HHH} \mid \text{Fair coin})P(\text{Fair coin})}$$

$$= \frac{1 \times \frac{1}{3}}{1 \times \frac{1}{3} + \frac{1}{8} \times \frac{2}{3}} = \frac{\frac{1}{3}}{\frac{10}{24}} = \frac{4}{5}$$

Drug Testing

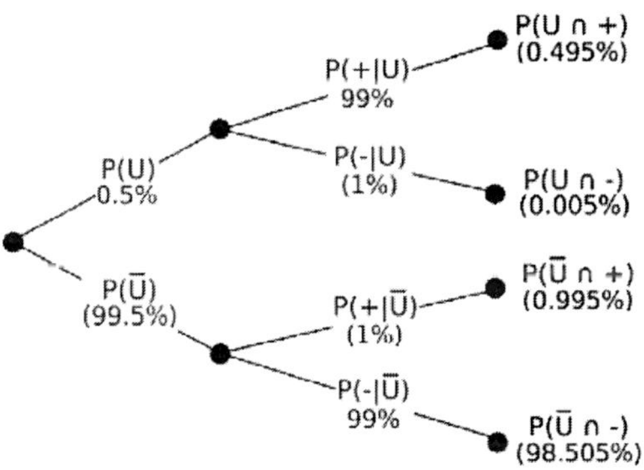

Figure: *Tree diagram illustrating drug testing example. U, U bar, "+" and "" are the events representing user, non-user, positive result and negative result. Percentages in parentheses are calculated.*

Suppose a drug test is 99% sensitive and 99% specific. That is, the test will produce 99% true positive results for drug users and 99% true negative results for non-drug users. Suppose that 0.5% of people are users of the drug. If a randomly selected individual tests positive, what is the probability he or she is a user?

$$P(\text{User} \mid +) = \frac{P(+ \mid \text{User})P(\text{User})}{P(+ \mid \text{User})P(\text{User}) + P(+ \mid \text{Non-user})P(\text{Non-user})}$$

$$= \frac{0.99 \times 0.005}{0.99 \times 0.005 + 0.01 \times 0.995}$$

$$\approx 33.2\%$$

Despite the apparent accuracy of the test, if an individual tests positive, it is more likely that they do *not* use the drug than that they do.

This surprising result arises because the number of non-users is very large compared to the number of users; thus the number of false positives (0.995%) outweighs the number of true positives (0.495%). To use concrete numbers, if 1000 individuals are tested, there are expected to be 995 non-users and 5 users. From the 995 non-users, $0.01 \times 995 \simeq 10$ false positives are expected. From the 5 users, $0.99 \times 5 \simeq 5$ true positives are expected. Out of 15 positive results, only 5, about 33%, are genuine.

Note: The importance of specificity can be illustrated by showing that even if sensitivity is 100% and specificity is at 99% the probability of the person being a drug user is H"33% but if the specificity is changed to 99.5% and the sensitivity is dropped down to 99% the probability of the person being a drug user rises to 49.8%.

Applications

Bayes's theorem is significantly important in inverse problem theory, where the *a posteriori* probability density function is obtained from the product of prior probability density function and the likelihood probability density function. An important application is constructing computational models of oil reservoirs given the observed data. Although Bayes's theorem is commonly used to determine the probability of an event occurring, it can also be applied to verify someone's credibility as a prognosticator. Many pundits claim to be able to predict the outcome of an event; political elections, trials, the weather and even sporting events. Larry Sabato, founder of Sabato's Crystal Ball, is a perfect example. His website provides free political analysis and election predictions. His success at predictions has even led him to be called a "pundit with an opinion for every reporter's phone call." We even have Punxsutawney Phil, the famous groundhog, who tells us whether or not we can expect a longer winter or an early spring. Bayes's theorem tells us the difference between who's on a hot streak and who is what they claim to be.

Let's say we live in an area where everyone gambles on the outcome of coin flips. Because of that, there is a big business for predicting coin flips. Suppose that 5% of predictors can actually win in the long run, and 80% of those are winners over a 2-year period. 95% of predictors are pretenders who are just guessing, and 20% of them are winners over a 2-year period (everyone gets lucky once in a while). This means that 82.6% of bettors who are winners over a 2-year period are actually long-term losers who are winning above their real average.

History

Bayes's theorem was named after the Reverend Thomas Bayes (1701–61), who studied how to compute a distribution for the probability parameter of a binomial distribution (in modern terminology). His friend Richard Price edited and presented this work in 1763, after Bayes's death, as *An Essay towards solving a Problem in the Doctrine of Chances*.

The French mathematician Pierre-Simon Laplace reproduced and extended Bayes's results in 1774, apparently quite unaware of Bayes's work. Stephen Stigler suggested in 1983 that Bayes's theorem was discovered by Nicholas Saunderson some time before Bayes. However, this interpretation has been disputed.

Martyn Hooper and Sharon McGrayne have argued that Richard Price's contribution was substantial:

By modern standards, we should refer to the Bayes–Price rule. Price discovered Bayes's work, recognised its importance, corrected it, contributed to the article, and found a use for it. The modern convention of employing Bayes's name alone is unfair but so entrenched that anything else makes little sense.

Boole's Inequality

In probability theory, Boole's inequality, also known as the union bound, says that for any finite or countable set of events, the probability that at least one of the events happens is no greater than the sum of the probabilities of the individual events. Boole's inequality is named after George Boole.

Formally, for a countable set of events A_1, A_2, A_3, ..., we have

$$\mathbb{P}\left(\bigcup_i A_i\right) \le \sum_i \mathbb{P}(A_i).$$

In measure-theoretic terms, Boole's inequality follows from the fact that a measure (and certainly any probability measure) is σ-sub-additive.

Proof

Boole's inequality may be proved for finite collections of events using the method of induction.

For the $n = 1$ case, it follows that

$$\mathbb{P}(A_1) \le \mathbb{P}(A_1).$$

For the case n, we have

$$\mathbb{P}\Big(\bigcup_{i=1}^{n} A_i\Big) \le \sum_{i=1}^{n} \mathbb{P}(A_i).$$

Since $\mathbb{P}(A \cup B) = \mathbb{P}(A) + \mathbb{P}(B) - \mathbb{P}(A \cap B)$, and because the union operation is associative, we have

$$\mathbb{P}\Big(\bigcup_{i=1}^{n+1} A_i\Big) = \mathbb{P}\Big(\bigcup_{i=1}^{n} A_i\Big) + \mathbb{P}(A_{n+1}) - \mathbb{P}\Big(\bigcup_{i=1}^{n} A_i \cap A_{n+1}\Big).$$

Since

$$\mathbb{P}\Big(\bigcup_{i=1}^{n} A_i \cap A_{n+1}\Big) \ge 0,$$

by the first axiom of probability, we have

$$\mathbb{P}\Big(\bigcup_{i=1}^{n+1} A_i\Big) \le \mathbb{P}\Big(\bigcup_{i=1}^{n} A_i\Big) + \mathbb{P}(A_{n+1}),$$

and therefore

Bonferroni Inequalities

Boole's inequality may be generalised to find upper and lower bounds on the probability of finite unions of events. These bounds are known as Bonferroni inequalities, after Carlo Emilio Bonferroni.

Define

$$S_1 := \sum_{i=1}^{n} \mathbb{P}(A_i),$$

and

$$S_2 := \sum_{1 \le i < j \le n} \mathbb{P}(A_i \cap A_j),$$

as well as

$$S_k := \sum_{1 \le i_1 < \cdots < i_k \le n} \mathbb{P}(A_{i_1} \cap \cdots \cap A_{i_k})$$

for all integers k in $\{3, \ldots, n\}$.

Then, for odd k in $\{1, \ldots, n\}$,

$$\mathbb{P}\Big(\bigcup_{i=1}^{n} A_i\Big) \le \sum_{j=1}^{k} (-1)^{j-1} S_j,$$

and for even k in $\{2, \ldots, n\}$,

$$\mathbb{P}\Big(\bigcup_{i=1}^{n} A_i\Big) \ge \sum_{j=1}^{k} (-1)^{j-1} S_j.$$

Boole's inequality is recovered by setting $k = 1$. When $k = n$, then equality holds and the resulting identity is the inclusion–exclusion principle.

Law of Large Numbers

In probability theory, the law of large numbers (LLN) is a theorem that describes the result of performing the same experiment a large number of times. According to the law, the average of the results obtained from a large number of trials should be close to the expected value, and will tend to become closer as more trials are performed.

The LLN is important because it "guarantees" stable long-term results for the averages of some random events. For example, while a casino may lose money in a single spin of the roulette wheel, its earnings will tend towards a predictable percentage over a large number of spins. Any winning streak by a player will eventually be overcome by the parameters of the game. It is important to remember that the LLN only applies (as the name indicates) when a *large number* of observations are considered. There is no principle that a small number of observations will coincide with the expected value or that a streak of one value will immediately be "balanced" by the others.

Examples

For example, a single roll of a fair six-sided die produces one of the numbers 1, 2, 3, 4, 5, or 6, each with equal probability. Therefore, the expected value of a single die roll is

$$\frac{1+2+3+4+5+6}{6} = 3.5.$$

According to the law of large numbers, if a large number of six-sided dice are rolled, the average of their values (sometimes called the sample mean) is likely to be close to 3.5, with the precision increasing as more dice are rolled.

It follows from the law of large numbers that the empirical probability of success in a series of Bernoulli trials will converge to the theoretical probability. For a Bernoulli random variable, the expected value is the theoretical probability of success, and the average of n such variables (assuming they are independent and identically distributed (i.i.d.)) is precisely the relative frequency.

For example, a fair coin toss is a Bernoulli trial. When a fair coin is flipped once, the theoretical probability that the outcome will be heads is equal to 1/2. Therefore, according to the law of large numbers, the proportion of heads in a "large" number of coin flips "should be" roughly 1/2. In particular, the proportion of heads after n flips will almost surely converge to 1/2 as n approaches infinity.

Though the proportion of heads (and tails) approaches 1/2, almost surely the absolute (nominal) difference in the number of heads and tails will become large as the number of flips becomes large. That is, the probability that the absolute difference is a small number, approaches zero as the number of flips becomes large. Also, almost surely the ratio of the absolute difference to the number of flips will approach zero. Intuitively, expected absolute difference grows, but at a slower rate than the number of flips, as the number of flips grows.

History

The Italian mathematician Gerolamo Cardano (1501–1576) stated without proof that the accuracies of empirical statistics tend to improve with the number of trials. This was then formalized as a law of large numbers. A special form of the LLN (for a binary random variable) was first proved by Jacob Bernoulli. It took him over 20 years to develop a sufficiently rigorous mathematical proof which was published in his *Ars Conjectandi* (The Art of Conjecturing) in 1713. He named this his "Golden Theorem" but it became generally known as "Bernoulli's Theorem". This should not be confused with the principle in physics with the same name, named after Jacob Bernoulli's nephew Daniel Bernoulli. In 1837, S.D. Poisson further described it under the name "la loi des grands nombres" ("The law of large numbers"). Thereafter, it was known under both names, but the "Law of large numbers" is most frequently used.

After Bernoulli and Poisson published their efforts, other mathematicians also contributed to refinement of the law, including Chebyshev, Markov, Borel, Cantelli and Kolmogorov and Khinchin, who finally provided a complete proof of the LLN for arbitrary random variables. These further studies have given rise to two prominent forms of the LLN. One is called the "weak" law and the other the "strong" law, in reference to two different modes of convergence of the cumulative sample means to the expected value; in particular, as explained below, the strong form implies the weak.

Forms

Two different versions of the law of large numbers are described below; they are called the *strong law of large numbers*, and the *weak law of large numbers*. Both versions of the law state that – with virtual certainty – the sample average

$$\bar{X}_n = \frac{1}{n}(X_1 + \cdots + X_n)$$

converges to the expected value

$$\bar{X}_n \to \mu \qquad \text{for} \qquad n \to \infty,$$

where X_1, X_2, ... is an infinite sequence of i.i.d. Lebesgue integrable random variables with expected value $E(X_1) = E(X_2) = ... = \mu$. Lebesgue integrability of X_j means that the expected value $E(X_j)$ exists according to Lebesgue integration and is finite.

An assumption of finite variance $Var(X_1) = Var(X_2) = ... = \sigma^2 < \infty$ is not necessary. Large or infinite variance will make the convergence slower, but the LLN holds anyway. This assumption is often used because it makes the proofs easier and shorter.

The difference between the strong and the weak version is concerned with the mode of convergence being asserted.

Weak Law

The weak law of large numbers (also called Khintchine's law) states that the sample average converges in probability towards the expected value

$$\bar{X}_n \xrightarrow{P} \mu \qquad \text{when } n \to \infty.$$

That is to say that for any positive number ε,

$$\lim_{n\to\infty} \Pr\left(|\bar{X}_n - \mu| > \varepsilon\right) = 0.$$

Interpreting this result, the weak law essentially states that for any nonzero margin specified, no matter how small, with a sufficiently large sample there will be a very high probability that the average of the observations will be close to the expected value; that is, within the margin.

Convergence in probability is also called weak convergence of random variables. This version is called the weak law because random variables may converge weakly (in probability) as above without converging strongly (almost surely) as below.

Strong Law

The strong law of large numbers states that the sample average converges almost surely to the expected value

$$\bar{X}_n \xrightarrow{a.s.} \mu \qquad \text{when } n \to \infty.$$

That is,

$$\Pr\left(\lim_{n\to\infty} \overline{X}_n = \mu\right) = 1.$$

The proof is more complex than that of the weak law. This law justifies the intuitive interpretation of the expected value (for Lebesgue integration only) of a random variable when sampled repeatedly as the "long-term average".

Almost sure convergence is also called strong convergence of random variables. This version is called the strong law because random variables which converge strongly (almost surely) are guaranteed to converge weakly (in probability). The strong law implies the weak law.

The strong law of large numbers can itself be seen as a special case of the pointwise ergodic theorem.

Moreover, if the summands are independent but not identically distributed, then

$$\overline{X}_n - \mathrm{E}\left[\overline{X}_n\right] \xrightarrow{a.s.} 0,$$

provided that each X_k has a finite second moment and

$$\sum_{k=1}^{\infty} \frac{1}{k^2} \mathrm{Var}[X_k] < \infty.$$

This statement is known as *Kolmogorov's strong law.*

Differences between the Weak Law and the Strong Law

The *weak law* states that for a specified large *n*, the average $\overline{X}_n$ is likely to be near μ. Thus, it leaves open the possibility that $|\overline{X}_n - \mu| > \varepsilon$ happens an infinite number of times, although at infrequent intervals.

The *strong law* shows that this almost surely will not occur. In particular, it implies that with probability 1, we have that for any ε > 0 the inequality $|\overline{X}_n - \mu| < \varepsilon$ holds for all large enough *n*.

The strong law does not hold in the following cases, but the weak law does

1. Let x be exponentially distributed random variable with parameter 1, the transformation with the following expected value: $E\left(\frac{\sin(x)e^x}{x}\right) = \int_0^\infty \frac{\sin(x)e^x}{x} e^{-x} dx = \frac{\pi}{2}$
2. Let x be geometric distribution with probability 0.5, the transformation with the following expected value:

$$E\left(\frac{2^x(-1)^x}{x}\right) = \sum_1^{\infty} \frac{2^x(-1)^x}{x} 2^{-x} = -\log(2)$$

3. $$1 - F(x) = \frac{e}{2x\log(x)}, x \geq e$$

$$F(x) = \frac{e}{-2x\log(-x)}, x \leq -e$$

Uniform Law of Large Numbers

Suppose $f(x,\theta)$ is some function defined for $\theta \in \Theta$, and continuous in θ. Then for any fixed θ, the sequence $\{f(X_1,\theta), f(X_2,\theta), \ldots\}$ will be a sequence of independent and identically distributed random variables, such that the sample mean of this sequence converges in probability to $E[f(X,\theta)]$. This is the *pointwise* (in θ) convergence.

The uniform law of large numbers states the conditions under which the convergence happens *uniformly* in θ. If

1. Θ is compact,
2. $f(x,\theta)$ is continuous at each $\theta \in \Theta$ for almost all x's, and measurable function of x at each θ.
3. there exists a dominating function $d(x)$ such that $E[d(X)] < \infty$, and

$$\|f(x,\theta)\| \leq d(x) \quad \text{for all } \theta \in \Theta.$$

Then $E[f(X,\theta)]$ is continuous in θ, and

$$\sup_{\theta \in \Theta} \left\| \frac{1}{n} \sum_{i=1}^{n} f(X_i,\theta) - E[f(X,\theta)] \right\| \xrightarrow{\text{a.s.}} 0.$$

Borel's Law of Large Numbers

Borel's law of large numbers, named after Émile Borel, states that if an experiment is repeated a large number of times, independently under identical conditions, then the proportion of times that any specified event occurs approximately equals the probability of the event's occurrence on any particular trial; the larger the number of repetitions, the better the approximation tends to be. More precisely, if E denotes the event in question, p its probability of occurrence, and $N_n(E)$ the number of times E occurs in the first n trials, then with probability one,

$$\frac{N_n(E)}{n} \to p \text{ as } n \to \infty.$$

Chebyshev's Lemma. Let X be a random variable with finite expected value μ and finite non-zero variance σ^2. Then for any real number $k > 0$,

$$\Pr(|X - \mu| \geq k\sigma) \leq \frac{1}{k^2}.$$

This theorem makes rigorous the intuitive notion of probability as the long-run relative frequency of an event's occurrence. It is a special case of any of several more general laws of large numbers in probability theory.

Proof

Given $X_1, X_2, \ldots$ an infinite sequence of i.i.d. random variables with finite expected value $E(X_1) = E(X_2) = \ldots = \mu < \infty$, we are interested in the convergence of the sample average

$$\overline{X}_n = \tfrac{1}{n}(X_1 + \cdots + X_n).$$

The weak law of large numbers states:

Theorem: $\overline{X}_n \xrightarrow{P} \mu \quad \text{for} \quad n \to \infty.$

Proof Using Chebyshev's Inequality

This proof uses the assumption of finite variance $\text{Var}(X_i) = \sigma^2$ (for all i). The independence of the random variables implies no correlation between them, and we have that

$$\text{Var}(\overline{X}_n) = \text{Var}(\tfrac{1}{n}(X_1 + \cdots + X_n)) = \frac{1}{n^2}\text{Var}(X_1 + \cdots + X_n) = \frac{n\sigma^2}{n^2} = \frac{\sigma^2}{n}.$$

The common mean μ of the sequence is the mean of the sample average:

$$E(\overline{X}_n) = \mu.$$

Using Chebyshev's inequality on $\overline{X}_n$ results in

$$\text{P}(\left|\overline{X}_n - \mu\right| \geq \varepsilon) \leq \frac{\sigma^2}{n\varepsilon^2}.$$

This may be used to obtain the following:

$$\text{P}(\left|\overline{X}_n - \mu\right| < \varepsilon) = 1 - \text{P}(\left|\overline{X}_n - \mu\right| \geq \varepsilon) \geq 1 - \frac{\sigma^2}{n\varepsilon^2}.$$

As n approaches infinity, the expression approaches 1. And by definition of convergence in probability, we have obtained

$$\overline{X}_n \xrightarrow{P} \mu \qquad \text{for} \qquad n \to \infty.$$

Proof Using Convergence of Characteristic Functions

By Taylor's theorem for complex functions, the characteristic function of any random variable, X, with finite mean μ, can be written as

$$\varphi_X(t) = 1 + it\mu + o(t), \quad t \to 0.$$

All $X_1, X_2, \ldots$ have the same characteristic function, so we will simply denote this φ_X.

Among the basic properties of characteristic functions there are

$$\varphi_{\frac{1}{n}X}(t) = \varphi_X(\tfrac{t}{n}) \quad \text{and} \quad \varphi_{X+Y}(t) = \varphi_X(t)\varphi_Y(t)$$

if X and Y are independent.

These rules can be used to calculate the characteristic function of $\overline{X}_n$ in terms of φ_X:

$$\varphi_{\overline{X}_n}(t) = \left[\varphi_X\left(\frac{t}{n}\right)\right]^n = \left[1 + i\mu\frac{t}{n} + o\left(\frac{t}{n}\right)\right]^n \to e^{it\mu}, \quad \text{as} \quad n \to \infty.$$

The limit $e^{it\mu}$ is the characteristic function of the constant random variable μ, and hence by the Lévy continuity theorem, $\overline{X}_n$ converges in distribution to μ:

$$\overline{X}_n \xrightarrow{\mathcal{D}} \mu \qquad \text{for} \qquad n \to \infty.$$

μ is a constant, which implies that convergence in distribution to μ and convergence in probability to μ are equivalent. Therefore,

$$\overline{X}_n \xrightarrow{P} \mu \qquad \text{for} \qquad n \to \infty.$$

This shows that the sample mean converges in probability to the derivative of the characteristic function at the origin, as long as the latter exists.

Marginal Distribution

In probability theory and statistics, the marginal distribution of a subset of a collection of random variables is the probability distribution of the variables contained in the subset. It gives the probabilities of various values of the variables in the subset without reference to the values of the other variables. This contrasts with a conditional distribution, which gives the probabilities contingent upon the values of the other variables.

The term marginal variable is used to refer to those variables in the subset of variables being retained. These terms are dubbed "marginal" because they used to be found by summing values in a table along rows or columns, and writing the sum in the margins of the table. The distribution of the marginal variables (the marginal distribution) is obtained by marginalizing over the distribution of the variables being discarded, and the discarded variables are said to have been marginalized out.

The context here is that the theoretical studies being undertaken, or the data analysis being done, involves a wider set of random variables but that attention is being limited to a reduced number of those variables. In many applications an analysis may start with a given collection of random variables, then first extend the set by defining new ones (such as the sum of the original random variables) and finally reduce the number by placing interest in the marginal distribution of a subset (such as the sum). Several different analyses may be done, each treating a different subset of variables as the marginal variables.

Two-Variable Case

	x_1	x_2	x_3	x_4	$p_y(Y)$ ↓
y_1	$4/32$	$2/32$	$1/32$	$1/32$	$8/32$
y_2	$2/32$	$4/32$	$1/32$	$1/32$	$8/32$
y_3	$2/32$	$2/32$	$2/32$	$2/32$	$8/32$
y_4	$8/32$	0	0	0	$8/32$
$p_x(X)$ →	$16/32$	$8/32$	$4/32$	$4/32$	$32/32$

Joint and marginal distributions of a pair of discrete, random variables X,Y having nonzero mutual information I(X; Y). The values of the joint distribution are in the 4×4 square, and the values of the marginal distributions are along the right and bottom margins.

Joint and marginal distributions of a pair of discrete, random variables X,Y having nonzero mutual information I(X; Y). The values of the joint distribution are in the 4×4 square, and the values of the marginal distributions are along the right and bottom margins.

Given two random variables X and Y whose joint distribution is known, the marginal distribution of X is simply the probability distribution of X averaging over information about Y. It is the probability distribution of X when the value of Y is not known. This is typically calculated by summing or integrating the joint probability distribution over Y.

For discrete random variables, the marginal probability mass function can be written as $\Pr(X = x)$. This is

$$\Pr(X = x) = \sum_y \Pr(X = x, Y = y) = \sum_y \Pr(X = x \mid Y = y)\Pr(Y = y),$$

where $\Pr(X = x, Y = y)$ is the joint distribution of X and Y, while $\Pr(X = x \mid Y = y)$ is the conditional distribution of X given Y. In this case, the variable Y has been marginalized out.

Bivariate marginal and joint probabilities for discrete random variables are often displayed as two-way tables.

Similarly for continuous random variables, the marginal probability density function can be written as $p_X(x)$. This is

$$p_X(x) = \int_y p_{X,Y}(x,y)\mathrm{d}y = \int_y p_{X|Y}(x \mid y)p_Y(y)\mathrm{d}y,$$

where $p_{X,Y}(x,y)$ gives the joint distribution of X and Y, while $p_{X|Y}(x \mid y)$ gives the conditional distribution for X given Y. Again, the variable Y has been marginalized out.

Note that a marginal probability can always be written as an expected value:

$$p_X(x) = \int_y p_{X|Y}(x \mid y)p_Y(y)\mathrm{d}y = \mathbb{E}_Y[p_{X|Y}(x \mid y)]$$

Intuitively, the marginal probability of X is computed by examining the conditional probability of X given a particular value of Y, and then averaging this conditional probability over the distribution of all values of Y.

This follows from the definition of expected value, i.e. in general

$$\mathbb{E}_Y[f(Y)] = \int_y f(y)p_Y(y)\mathrm{d}y$$

Real-World Example

Suppose that the probability that a pedestrian will be hit by a car while crossing the road at a pedestrian crossing without paying attention to the traffic light is to be computed. Let H be a discrete random variable taking one value from {Hit, Not Hit}. Let L be a discrete random variable taking one value from {Red, Yellow, Green}.

Realistically, H will be dependent on L. That is, P(H = Hit) and P(H = Not Hit) will take different values depending on whether L is red, yellow or green. A person is, for example, far more likely to be hit by a car when trying to cross while the lights for cross traffic are green than if they are red. In other words, for any given possible pair of values for H and L, one must consider the joint probability distribution of H and L to find the probability of that pair of events occurring together if the pedestrian ignores the state of the light.

However, in trying to calculate the marginal probability P(H=hit), what we are asking for is the probability that H=Hit in the situation in which we don't actually know the particular value of L and in which the pedestrian ignores the state of the light. In general a pedestrian can be hit if the lights are red OR if the lights are yellow OR if the lights are green. So in this case the answer for the marginal probability can be found by summing P(H,L) for all possible values of L, with each value of L weighted by its probability of occurring.

Here is a table showing the conditional probabilities of being hit, depending on the state of the lights. (Note that the columns in this table must add up to 1 because the probability of being hit or not hit is 1 regardless of the state of the light.)

Conditional distribution: P(H|L)			
	L=Red	L=Yellow	L=Green
H=Not Hit	0.99	0.9	0.2
H=Hit	0.01	0.1	0.8

To find the joint probability distribution, we need more data. Let's say that P(L=red) = 0.2, P(L=yellow) = 0.1, and P(L=green) = 0.7. Multiplying each column in the conditional distribution by the probability of that column occurring, we find the joint probability distribution of H and L, given in the central 2×3 block of entries. (Note that the cells in this 2×3 block add up to 1).

Joint distribution: P(H,L)				
	L=Red	***L=Yellow***	***L=Green***	***Marginal probability P(H)***
H=Not Hit	0.198	0.09	0.14	0.428
H=Hit	0.002	0.01	0.56	0.572
Total	0.2	0.1	0.7	1

The marginal probability P(H=Hit) is the sum along the H=Hit row of this joint distribution table, as this is the probability of being hit when the lights are red OR yellow OR green. Similarly, the marginal probability that P(H=Not Hit) is the sum of the H=Not Hit row.

Multivariate Distributions

For multivariate distributions, formulae similar to those above apply with the symbols X and/or Y being interpreted as vectors. In particular, each summation or integration would be over all variables except those contained in X.

Conditional Probability

In probability theory, a conditional probability measures the probability of an event given that (by assumption, presumption, assertion or evidence) another event has occurred. For example, the conditional probability of having a cold given that you are coughing might be 75%, meaning you probably have a cold if you are coughing. But the non-conditional probability (normally called just "probability") of having a cold may be only 5%, meaning that only 5% of the population as a whole has a cold, including people who are coughing and people who are not.

The concept of conditional probability is one of the most fundamental and one of the most important concepts in probability theory. But conditional probabilities can be quite slippery and require careful interpretation. For example, there need not be a causal or temporal relationship between A and B.

If the event of interest is A and the event B has already occurred, "the conditional probability of A given B", or "the probability of A under the condition B", is commonly denoted by $P(A \mid B)$, or sometimes $P_B(A)$.

In statistical inference, the conditional probability is an update of the probability of an event based on new information. Incorporating the new information can be done as follows

- Let A the event of interest be in the sample space, say (X,P).
- The occurrence of the event A knowing that event B has or will have occurred, means the occurrence of A as it is restricted to B, i.e. $A \cap B$.
- Without the knowledge of the occurence of B, the information about the occurence of A would simply be $P(A)$
- The probability of A knowing that event B has or will have occurred, will be the probability of compared with $P(B)$, the probability B has occured.

- This results in $P(A \mid B) = P(A\ B)/P(B)$ whenever $P(B)>0$ and 0 otherwise.

Note: This approach results in a probability measure that is consistent with the original probability measure and satisfies all the Kolmogorov Axioms. This conditional probability measure also could have resulted by assuming that the relative magnitude of the probability of A with respect to X will be preserved with respect to B (cf. a Formal Derivation below).

Note: The phraseology "evidence" or "information" is generally used in the Bayesian interpretation of probability. The conditioning event is interpreted as evidence for the conditioned event. That is, $P(A)$ is the probability of A before accounting for evidence E, and $P(A \mid E)$ is the probability of A after having accounted for evidence E or after having updated $P(A)$. This is consistent with the frequentist interpretation, which presumably is the first definition given above.

Note: $P(A \mid B)$ (the conditional probability of A given B) may or may not be equal to P(A) (the unconditional probability of A). If $P(A \mid B)$ = P(A), A and B are said to be independent.

Definition

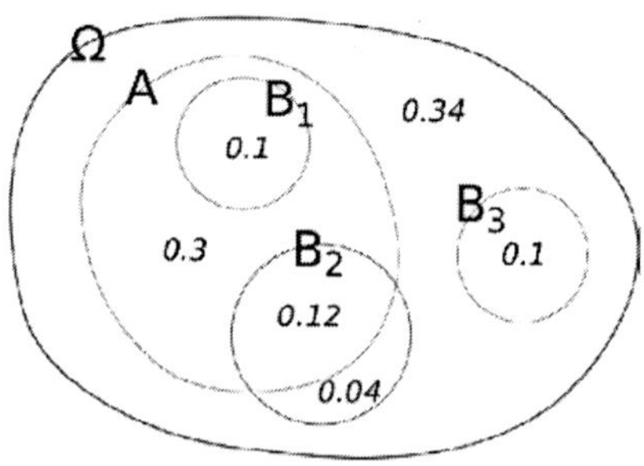

Figure: *Illustration of conditional probabilities with an Euler diagram. The unconditional probability* P(A) *= 0.52. However, the conditional probability* $P(A \mid B_1) = 1$, $P(A \mid B_2) \approx 0.75$, *and* $P(A \mid B_3) = 0$.

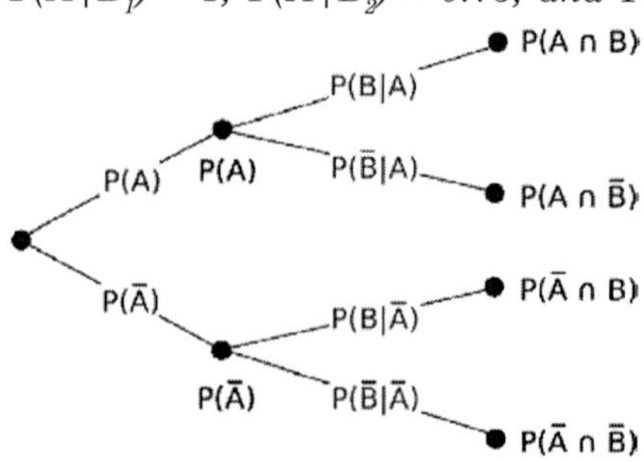

Figure: *On a tree diagram, branch probabilities are conditional on the event associated with the parent node.*

Kolmogorov Definition

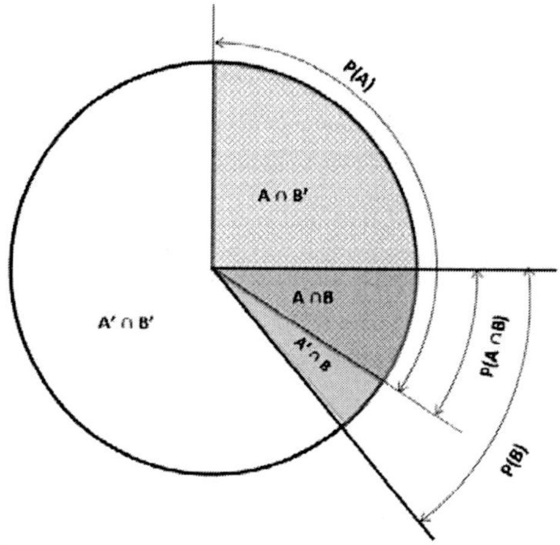

Figure: *Venn Pie Chart describing conditional probabilities*

Given two events A and B from the sigma-field of a probability space with $P(B) > 0$, the conditional probability of A given B is defined as the quotient of the probability of the joint of events A and B, and the probability of B:

$$P(A \mid B) = \frac{P(A \cap B)}{P(B)}$$

This may be visualized as restricting the sample space to B. The logic behind this equation is that if the outcomes are restricted to B, this set serves as the new sample space.

Note that this is a definition but not a theoretical result. We just denote the quantity $P(A \cap B)/P(B)$ as $P(A \mid B)$ and call it the conditional probability of A given B.

As an Axiom of Probability

Some authors, such as De Finetti, prefer to introduce conditional probability as an axiom of probability:

$$P(A \cap B) = P(A \mid B)P(B)$$

Although mathematically equivalent, this may be preferred philosophically; under major probability interpretations such as the subjective theory, conditional probability is considered a primitive entity. Further, this "multiplication axiom" introduces a symmetry with the summation axiom for mutually exclusive events:

$$P(A \cup B) = P(A) + P(B) - 0\cancel{P(A \cap B)}^{0}$$

Definition with σ-algebra

If $P(B) = 0$, then the simple definition of $P(A \mid B)$ is undefined. However, it is possible to define a conditional probability with respect to a σ-algebra of such events (such as those arising from a continuous random variable).

For example, if X and Y are non-degenerate and jointly continuous random variables with density $f_{X,Y}(x, y)$ then, if B has positive measure,

$$P(X \in A \mid Y \in B) = \frac{\int_{y \in B} \int_{x \in A} f_{X,Y}(x,y)\,dx\,dy}{\int_{y \in B} \int_{x \in \Omega} f_{X,Y}(x,y)\,dx\,dy}.$$

The case where B has zero measure can only be dealt with directly in the case that $B = \{y_0\}$, representing a single point, in which case

$$P(X \in A \mid Y = y_0) = \frac{\int_{x \in A} f_{X,Y}(x,y_0)\,dx}{\int_{x \in \Omega} f_{X,Y}(x,y_0)\,dx}.$$

If A has measure zero then the conditional probability is zero. An indication of why the more general case of zero measure cannot be dealt with in a similar way can be seen by noting that the limit, as all δy_i approach zero, of

$$P(X \in A \mid Y \in \cup_i [y_i, y_i + \delta y_i]) \approx \frac{\sum_i \int_{x \in A} f_{X,Y}(x,y_i)\,dx\,\delta y_i}{\sum_i \int_{x \in \Omega} f_{X,Y}(x,y_i)\,dx\,\delta y_i},$$

depends on their relationship as they approach zero.

Conditioning on a Random Variable

Conditioning on an event may be generalized to conditioning on a random variable. Let X be a random variable; we assume for the sake of presentation that X is discrete, that is, X takes on only finitely many values x. Let A be an event. The conditional probability of A given X is defined as the random variable, written $P(A \mid X)$, that takes on the value

$$P(A \mid X = x)$$

whenever

$$X = x.$$

More formally:

$$P(A \mid X)(\omega) = P(A \mid X = X(\omega)).$$

The conditional probability $P(A \mid X)$ is a function of X, e.g., if the function g is defined as

$$g(x) = P(A \mid X = x),$$

then

$$P(A \mid X) = g^{\circ}X$$

Note that $P(A \mid X)$ and X are now both random variables. From the law of total probability, the expected value of $P(A \mid X)$ is equal to the unconditional probability of A.

Example

Suppose that somebody secretly rolls two fair six-sided dice, and we must predict the outcome.

- Let A be the value rolled on die 1
- Let B be the value rolled on die 2

What is the probability that A = *2?*

Table 1 shows the sample space of 36 outcomes

Clearly, $A = 2$ in exactly 6 of the 36 outcomes, thus $P(A=2) = {}^{6}/_{36} = {}^{1}/_{6}$.

Table 1

+	B=1	2	3	4	5	6
A=1	2	3	4	5	6	7
2	3	4	5	6	7	8
3	4	5	6	7	8	9
4	5	6	7	8	9	10
5	6	7	8	9	10	11
6	7	8	9	10	11	12

Suppose it is revealed that A+B ≤ *5*

What is the probability $A+B \leq 5$?

Table 2 shows that $A+B \leq 5$ for exactly 10 of the same 36 outcomes, thus $P(A+B \leq 5) = {}^{10}/_{36}$

Table 2

+	B=1	2	3	4	5	6
A=1	2	3	4	5	6	7
2	3	4	5	6	7	8
3	4	5	6	7	8	9
4	5	6	7	8	9	10
5	6	7	8	9	10	11
6	7	8	9	10	11	12

What is the probability that A = *2* given that A+B ≤ *5 ?*

Table 3 shows that for 3 of these 10 outcomes, $A = 2$

Thus, the conditional probability $P(A=2 \mid A+B \leq 5) = {}^{3}/_{10} = 0.3$.

Table 3						
+	B=1	2	3	4	5	6
A=1	2	3	4	5	6	7
2	3	4	5	6	7	8
3	4	5	6	7	8	9
4	5	6	7	8	9	10
5	6	7	8	9	10	11
6	7	8	9	10	11	12

Statistical Independence

Events A and B are defined to be statistically independent if:

$$P(A \cap B) = P(A)P(B)$$

$$\Leftrightarrow P(A \mid B) = P(A)$$

$$\Leftrightarrow P(B \mid A) = P(B).$$

That is, the occurrence of A does not affect the probability of B, and vice versa. Although the derived forms may seem more intuitive, they are not the preferred definition as the conditional probabilities may be undefined if $P(A)$ or $P(B)$ are 0, and the preferred definition is symmetrical in A and B.

Common Fallacies

These fallacies should not be confused with Robert K. Shope's 1978 "conditional fallacy", which deals with counterfactual examples that beg the question.

Assuming Conditional Probability is of Similar Size to its Inverse

In general, it cannot be assumed that $P(A \mid B) \approx P(B \mid A)$. This can be an insidious error, even for those who are highly conversant with statistics.

The relationship between $P(A \mid B)$ and $P(B \mid A)$ is given by Bayes' theorem:

$$P(B \mid A) = \frac{P(A \mid B)P(B)}{P(A)}.$$

That is, $P(A \mid B) \approx P(B \mid A)$ only if $P(B)/P(A) \approx 1$, or equivalently, $P(A) \approx P(B)$.

Alternatively, noting that $A \cap B = B \cap A$, and applying conditional probability:

$$P(A \mid B)P(B) = P(A \cap B) = P(B \cap A) = P(B \mid A)P(A)$$

Rearranging gives the result.

Assuming Marginal and Conditional Probabilities are of Similar Size

In general, it cannot be assumed that $P(A) \approx P(A \mid B)$. These probabilities are linked through the law of total probability:

$$P(A) = \sum_n P(A \cap B_n) = \sum_n P(A \mid B_n)P(B_n).$$

where the events (B_n) form a countable partition of A.

This fallacy may arise through selection bias. For example, in the context of a medical claim, let S_C be the event that a sequela (chronic disease) S occurs as a consequence of circumstance (acute condition) C. Let H be the event that an individual seeks medical help. Suppose that in most cases, C does not cause S so $P(S_C)$ is low. Suppose also that medical attention is only sought if S has occurred due to C. From experience of patients, a doctor may therefore erroneously conclude that $P(S_C)$ is high. The actual probability observed by the doctor is $P(S_C \mid H)$.

Over- or Under-Weighting Priors

Not taking prior probability into account partially or completely is called *base rate neglect*. The reverse, insufficient adjustment from the prior probability is *conservatism*.

Formal Derivation

Formally, $P(A \mid B)$ is defined as the probability of A according to a new probability function on the sample space, such that outcomes not in B have probability 0 and that it is consistent with all original probability measures.

Let Ω be a sample space with elementary events $\{\Omega\}$. Suppose we are told the event $B \subseteq \Omega$ has occurred. A new probability distribution (denoted by the conditional notation) is to be assigned on $\{\omega\}$ to reflect this. For events in B, it is reasonable to assume that the relative magnitudes of the probabilities will be preserved. For some constant scale factor α, the new distribution will therefore satisfy:

1. $\omega \in B : P(\omega \mid B) = \alpha P(\omega)$
2. $\omega \notin B : P(\omega \mid B) = 0$
3. $\sum_{\omega \in \Omega} P(\omega \mid B) = 1.$

Substituting 1 and 2 into 3 to select α:

$$\begin{aligned}\sum_{\omega\in\Omega} P(\omega|B) &= \sum_{\omega\in B}\alpha P(\omega) + 0\cancel{\sum_{\omega\notin B} 0}\\ &= \alpha\sum_{\omega\in B} P(\omega)\\ &= \alpha\cdot P(B)\end{aligned}$$

$$\Rightarrow \alpha = \frac{1}{P(B)}$$

So the new probability distribution is

$$1.\ \omega\in B : P(\omega|B) = \frac{P(\omega)}{P(B)}$$

$$2.\ \omega\notin B : P(\omega|B) = 0$$

Now for a general event *A*,

$$\begin{aligned}P(A|B) &= \sum_{\omega\in A\cap B} P(\omega|B) + 0\cancel{\sum_{\omega\in A\cap B^c} P(\omega|B)}\\ &= \sum_{\omega\in A\cap B}\frac{P(\omega)}{P(B)}\\ &= \frac{P(A\cap B)}{P(B)}\end{aligned}$$

Complementary Event

In the study of probability, given at least two random variables *X*, *Y*, ..., that are defined on a probability space, the joint probability distribution for *X*, *Y*, ... is a probability distribution that gives the probability that each of *X*, *Y*, ... falls in any particular range or discrete set of values specified for that variable. In the case of only two random variables, this is called a bivariate distribution, but the concept generalizes to any number of random variables, giving a multivariate distribution.

The joint probability distribution can be expressed either in terms of a joint cumulative distribution function or in terms of a joint probability density function (in the case of continuous variables) or joint probability mass function (in the case of discrete variables). These in turn can be used to find two other types of distributions: the marginal distribution giving the probabilities for any one of the variables with no reference to any specific ranges of values for the other variables, and the conditional probability distribution giving the probabilities for any subset of the variables conditional on particular values of the remaining variables.

Example

Consider the roll of a fair dice and let $A = 1$ if the number is even (i.e. 2, 4, or 6) and $A = 0$ otherwise. Furthermore, let $B = 1$ if the number is prime (i.e. 2, 3, or 5) and $B = 0$ otherwise.

	1	2	3	4	5	6
A	0	1	0	1	0	1
B	0	1	1	0	1	0

Then, the joint distribution of A and B, expressed as a probability mass function, is

$$P(A=0,B=0)=P\{1\}=\frac{1}{6},\ P(A=1,B=0)=P\{4,6\}=\frac{2}{6},$$

$$P(A=0,B=1)=P\{3,5\}=\frac{2}{6},\ P(A=1,B=1)=P\{2\}=\frac{1}{6}.$$

These probabilities necessarily sum to 1, since the probability of *some* combination of A and B occurring is 1.

Important named Distributions

Named joint distributions that arise frequently in statistics include the multivariate normal distribution, the multivariate stable distribution, the multinomial distribution, the negative multinomial distribution, the multivariate hypergeometric distribution, and the elliptical distribution.

Cumulative Distribution

The joint probability distribution for a pair of random variables can be expressed in terms of their cumulative distribution function $F(x,y) = P(X \le x, Y \le y)$.

Density Function or Mass Function

Discrete Case

The joint probability mass function of two discrete random variables is equal to:

$$P(X = x \text{ and } Y = y) = P(Y = y \mid X = x) \cdot P(X = x) = P(X = x \mid Y = y) \cdot P(Y = y).$$

In general, the joint probability distribution of n discrete random variables $X_1, X_2, \ldots, X_n$ is equal to:

$$\begin{aligned}P(X_1 = x_1,\dots,X_n = x_n) &= P(X_1 = x_1)\\ &\quad\times P(X_2 = x_2 \mid X_1 = x_1)\\ &\quad\quad\times P(X_3 = x_3 \mid X_1 = x_1, X_2 = x_2)\times\dots\times P(X_n = x_n \mid X_1 = x_1, X_2 = x_2,\dots,X_{n-1} = x_{n-1})\end{aligned}$$

This identity is known as the chain rule of probability.

Since these are probabilities, we have in the two-variable case

$$\sum_i \sum_j P(X = x_i \text{ and } Y = y_j) = 1,$$

which generalizes for n discrete random variables $X_1, X_2, \dots, X_n$ to

$$\sum_i \sum_j \dots \sum_k P(X_1 = x_{1i}, X_2 = x_{2j}, \dots, X_n = x_{nk}) = 1.$$

Continuous Case

The joint probability density function $f_{X,Y}(x, y)$ for continuous random variables is equal to:

$$f_{X,Y}(x,y) = f_{Y|X}(y \mid x) f_X(x) = f_{X|Y}(x \mid y) f_Y(y)$$

...where $f_{Y|X}(y \mid x)$ and $f_{X|Y}(x \mid y)$ give the conditional distributions of Y given $X = x$ and of X given $Y = y$ respectively, and $f_X(x)$ and $f_Y(y)$ give the marginal distributions for X and Y respectively.

Again, since these are probability distributions, one has

$$\int_x \int_y f_{X,Y}(x,y)\, dy\, dx = 1.$$

Mixed Case

The "mixed joint density" may be defined where one random variable X is continuous and the other random variable Y is discrete, or vice versa, as:

$$f_{X,Y}(x,y) = f_{X|Y}(x \mid y) P(Y = y) = P(Y = y \mid X = x) f_X(x)$$

One example of a situation in which one may wish to find the cumulative distribution of one random variable which is continuous and another random variable which is discrete arises when one wishes to use a logistic regression in predicting the probability of a binary outcome Y conditional on the value of a continuously distributed outcome X. One *must* use the "mixed" joint density when finding the cumulative distribution of this binary outcome because the input variables (X, Y) were initially defined in such a way that one could not collectively assign it either a probability density function or a probability mass function. Formally, $f_{X,Y}(x, y)$ is the probability density function of (X, Y) with respect to the product measure on the respective supports of X and Y. Either of these two decompositions can then be used to recover the joint cumulative distribution function:

$$F_{X,Y}(x,y) = \sum_{t \le y} \int_{s=-\infty}^{x} f_{X,Y}(s,t)\, ds$$

The definition generalizes to a mixture of arbitrary numbers of discrete and continuous random variables.

Joint Distribution for Independent Variables

If for discrete random variables $P(X = x \text{ and } Y = y) = P(X = x) \cdot P(Y = y)$ for all x and y, or for absolutely continuous random variables $f_{X,Y}(x,y) = f_X(x) \cdot f_Y(y)$ for all x and y, then X and Y are said to be independent. This means that acquiring any information about the value of one or more of the random variables leads to a conditional distribution of any other variable that is identical to its unconditional (marginal) distribution; thus no variable provides any information about any other variable.

Joint Distribution for Conditionally Dependent Variables

If a subset A of the variables $X_1, \cdots, X_n$ is conditionally dependent given another subset B of these variables, then the joint distribution $\mathrm{P}(X_1, \ldots, X_n)$ is equal to $P(B) \cdot P(A \mid B)$. Therefore, it can be efficiently represented by the lower-dimensional probability distributions $P(B)$ and $P(A \mid B)$. Such conditional independence relations can be represented with a Bayesian network.

Sampling (Statistics)

In statistics, quality assurance, & survey methodology, sampling is concerned with the selection of a subset of individuals from within a statistical population to estimate characteristics of the whole population. Each observation measures one or more properties (such as weight, location, colour) of observable bodies distinguished as independent objects or individuals. In survey sampling, weights can be applied to the data to adjust for the sample design, particularly stratified sampling. Results from probability theory and statistical theory are employed to guide practice. In business and medical research, sampling is widely used for gathering information about a population.

The sampling process comprises several stages:

- Defining the population of concern
- Specifying a sampling frame, a set of items or events possible to measure

- Specifying a sampling method for selecting items or events from the frame
- Determining the sample size
- Implementing the sampling plan
- Sampling and data collecting
- Data which can be selected

Population Definition

Successful statistical practice is based on focused problem definition. In sampling, this includes defining the population from which our sample is drawn. A population can be defined as including all people or items with the characteristic one wishes to understand. Because there is very rarely enough time or money to gather information from everyone or everything in a population, the goal becomes finding a representative sample (or subset) of that population.

Sometimes what defines a population is obvious. For example, a manufacturer needs to decide whether a batch of material from production is of high enough quality to be released to the customer, or should be sentenced for scrap or rework due to poor quality. In this case, the batch is the population. Although the population of interest often consists of physical objects, sometimes we need to sample over time, space, or some combination of these dimensions. For instance, an investigation of supermarket staffing could examine checkout line length at various times, or a study on endangered penguins might aim to understand their usage of various hunting grounds over time. For the time dimension, the focus may be on periods or discrete occasions.

In other cases, our 'population' may be even less tangible. For example, Joseph Jagger studied the behaviour of roulette wheels at a casino in Monte Carlo, and used this to identify a biased wheel. In this case, the 'population' Jagger wanted to investigate was the overall behaviour of the wheel (i.e. the probability distribution of its results over infinitely many trials), while his 'sample' was formed from observed results from that wheel. Similar considerations arise when taking repeated measurements of some physical characteristic such as the electrical conductivity of copper. This situation often arises when we seek knowledge about the cause system of which the *observed* population is an outcome. In such cases, sampling theory may treat the observed population as a sample from a larger 'superpopulation'. For example, a researcher might study the success rate of a new 'quit smoking' program on a test group of 100 patients, in order to predict the effects of the program if it were made available nationwide. Here the superpopulation is "everybody in the country, given access to this treatment" - a group which does not yet exist, since the program isn't yet available to all.

Note also that the population from which the sample is drawn may not be the same as the population about which we actually want information. Often there is large but not complete overlap between these two groups due to frame issues etc. Sometimes they may be entirely separate - for instance, we might study rats in order to get a better understanding of human health, or we might study records from people born in 2008 in order to make predictions about people born in 2009.

Time spent in making the sampled population and population of concern precise is often well spent, because it raises many issues, ambiguities and questions that would otherwise have been overlooked at this stage.

Sampling Frame

In the most straightforward case, such as the sentencing of a batch of material from production (acceptance sampling by lots), it is possible to identify and measure every single item in the population and to include any one of them in our sample. However, in the more general case this is not possible. There is no way to identify all rats in the set of all rats. Where voting is not compulsory, there is no way to identify which people will actually vote at a forthcoming election (in advance of the election). These imprecise populations are not amenable to sampling in any of the ways below and to which we could apply statistical theory.

As a remedy, we seek a sampling frame which has the property that we can identify every single element and include any in our sample. The most straightforward type of frame is a list of elements of the population (preferably the entire population) with appropriate contact information. For example, in an opinion poll, possible sampling frames include an electoral register and a telephone directory.

Probability and Nonprobability Sampling

Probability Sampling

A probability sample is a sample in which every unit in the population has a chance (greater than zero) of being selected in the sample, and this probability can be accurately determined. The combination of these traits makes it possible to produce unbiased estimates of population totals, by weighting sampled units according to their probability of selection.

Example: We want to estimate the total income of adults living in a given street. We visit each household in that street, identify all adults living there, and randomly select one adult from each household. (For example, we can allocate each person a random number, generated from a uniform distribution between 0 and 1, and select the person with the highest number in each household). We then interview the selected person and find their income.

People living on their own are certain to be selected, so we simply add their income to our estimate of the total. But a person living in a household of two adults has only a one-in-two chance of selection. To reflect this, when we come to such a household, we would count the selected person's income twice towards the total. (The person who is selected from that household can be loosely viewed as also representing the person who isn't selected.)

In the above example, not everybody has the same probability of selection; what makes it a probability sample is the fact that each person's probability is known. When every element in the population *does* have the same probability of selection, this is known as an 'equal probability of selection' (EPS) design. Such designs are also referred to as 'self-weighting' because all sampled units are given the same weight.

Probability sampling includes: Simple Random Sampling, Systematic Sampling, Stratified Sampling, Probability Proportional to Size Sampling, and Cluster or Multistage Sampling. These various ways of probability sampling have two things in common:

1. Every element has a known nonzero probability of being sampled and
2. involves random selection at some point.

Nonprobability Sampling

Nonprobability sampling is any sampling method where some elements of the population have *no* chance of selection (these are sometimes referred to as 'out of coverage'/'undercovered'), or where the probability of selection can't be accurately determined. It involves the selection of elements based on assumptions regarding the population of interest, which forms the criteria for selection. Hence, because the selection of elements is nonrandom, nonprobability sampling does not allow the estimation of sampling errors. These conditions give rise to exclusion bias, placing limits on how much information a sample can provide about the population. Information about the relationship between sample and population is limited, making it difficult to extrapolate from the sample to the population.

Example: We visit every household in a given street, and interview the first person to answer the door. In any household with more than one occupant, this is a nonprobability sample, because some people are more likely to answer the door (e.g. an unemployed person who spends most of their time at home is more likely to answer than an employed housemate who might be at work when the interviewer calls) and it's not practical to calculate these probabilities.

Nonprobability sampling methods include accidental sampling, quota sampling and purposive sampling. In addition, nonresponse effects may turn *any* probability design into a nonprobability design if the characteristics of nonresponse are not well understood, since nonresponse effectively modifies each element's probability of being sampled.

Sampling Methods

Within any of the types of frame identified above, a variety of sampling methods can be employed, individually or in combination. Factors commonly influencing the choice between these designs include:

- Nature and quality of the frame
- Availability of auxiliary information about units on the frame
- Accuracy requirements, and the need to measure accuracy
- Whether detailed analysis of the sample is expected
- Cost/operational concerns

Simple Random Sampling

In a simple random sample (SRS) of a given size, all such subsets of the frame are given an equal probability. Furthermore, any given *pair* of elements has the same chance of selection as any other such pair (and similarly for triples, and so on). This minimises bias and simplifies analysis of results. In particular, the variance between individual results within the sample is a good indicator of variance in the overall population, which makes it relatively easy to estimate the accuracy of results.

However, SRS can be vulnerable to sampling error because the randomness of the selection may result in a sample that doesn't reflect the makeup of the population. For instance, a simple random sample of ten people from a given country will *on average* produce five men and five women, but any given trial is likely to overrepresent one sex and underrepresent the other. (Systematic and stratified techniques), attempt to overcome this problem by "using information about the population" to choose a more "representative" sample.

SRS may also be cumbersome and tedious when sampling from an unusually large target population. In some cases, investigators are interested in "research questions specific" to subgroups of the population. For example, researchers might be interested in examining whether cognitive ability as a predictor of job performance is equally applicable across racial groups. SRS cannot accommodate the needs of researchers in this situation because it does not provide subsamples of the population. "Stratified sampling" addresses this weakness of SRS.

Systematic Sampling

Systematic sampling relies on arranging the study population according to some ordering scheme and then selecting elements at regular intervals through that ordered list. Systematic sampling involves a random start and then proceeds with the selection of every kth element from then onwards. In this case, k=(population size/sample size). It is important that the starting point is not automatically the first in the list, but is instead randomly chosen from within the first to the kth element in the list. A simple example would be to select every 10th name from the telephone directory (an 'every 10th' sample, also referred to as 'sampling with a skip of 10').

As long as the starting point is randomized, systematic sampling is a type of probability sampling. It is easy to implement and the stratification induced can make it efficient, *if* the variable by which the list is ordered is correlated with the variable of interest. 'Every 10th' sampling is especially useful for efficient sampling from databases.

For example, suppose we wish to sample people from a long street that starts in a poor area (house No. 1) and ends in an expensive district (house No. 1000). A simple random selection of addresses from this street could easily end up with too many from the high end and too few from the low end (or vice versa), leading to an unrepresentative sample. Selecting (e.g.) every 10th street number along the street ensures that the sample is spread evenly along the length of the street, representing all of these districts. (Note that if we always start at house #1 and end at #991, the sample is slightly biased towards the low end; by randomly selecting the start between #1 and #10, this bias is eliminated. However, systematic sampling is especially vulnerable to periodicities in the list. If periodicity is present and the period is a multiple or factor of the interval used, the sample is especially likely to be *un*representative of the overall population, making the scheme less accurate than simple random sampling.

For example, consider a street where the odd-numbered houses are all on the north (expensive) side of the road, and the even-numbered houses are all on the south (cheap) side. Under the sampling scheme given above, it is impossible to get a representative sample; either the houses sampled will *all* be from the odd-numbered, expensive side, or they will *all* be from the even-numbered, cheap side.

Another drawback of systematic sampling is that even in scenarios where it is more accurate than SRS, its theoretical properties make it difficult to *quantify* that accuracy. (In the two examples of systematic sampling that are given above, much of the potential sampling error is due to variation between neighbouring houses - but because this method never selects two neighbouring houses, the sample will not give us any information on that variation.)

As described above, systematic sampling is an EPS method, because all elements have the same probability of selection (in the example given, one in ten). It is *not* 'simple random sampling' because different subsets of the same size have different selection probabilities - e.g. the set {4,14,24,...,994} has a one-in-ten probability of selection, but the set {4,13,24,34,...} has zero probability of selection.

Systematic sampling can also be adapted to a non-EPS approach.

Stratified Sampling

Where the population embraces a number of distinct categories, the frame can be organised by these categories into separate "strata." Each stratum is then sampled as an independent sub-population, out of which individual elements can be randomly selected. There are several potential benefits to stratified sampling.

First, dividing the population into distinct, independent strata can enable researchers to draw inferences about specific subgroups that may be lost in a more generalized random sample.

Second, utilising a stratified sampling method can lead to more efficient statistical estimates (provided that strata are selected based upon relevance to the criterion in question, instead of availability of the samples). Even if a stratified sampling approach does not lead to increased statistical efficiency, such a tactic will not result in less efficiency than would simple random sampling, provided that each stratum is proportional to the group's size in the population.

Third, it is sometimes the case that data are more readily available for individual, pre-existing strata within a population than for the overall population; in such cases, using a stratified sampling approach may be more convenient than aggregating data across groups (though this may potentially be at odds with the previously noted importance of utilising criterion-relevant strata).

Finally, since each stratum is treated as an independent population, different sampling approaches can be applied to different strata, potentially enabling researchers to use the approach best suited (or most cost-effective) for each identified subgroup within the population.

There are, however, some potential drawbacks to using stratified sampling. First, identifying strata and implementing such an approach can increase the cost and complexity of sample selection, as well as leading to increased complexity of population estimates. Second, when examining multiple criteria, stratifying variables may be related to some, but not to others, further complicating the design, and potentially reducing the utility of the strata.

Finally, in some cases (such as designs with a large number of strata, or those with a specified minimum sample size per group), stratified sampling can potentially require a larger sample than would other methods (although in most cases, the required sample size would be no larger than would be required for simple random sampling.

A stratified sampling approach is most effective when three conditions are met

1. Variability within strata are minimized
2. Variability between strata are maximized
3. The variables upon which the population is stratified are strongly correlated with the desired dependent variable.

Advantages over Other Sampling Methods

1. Focuses on important subpopulations and ignores irrelevant ones.
2. Allows use of different sampling techniques for different subpopulations.
3. Improves the accuracy/efficiency of estimation.
4. Permits greater balancing of statistical power of tests of differences between strata by sampling equal numbers from strata varying widely in size.

Disadvantages

1. Requires selection of relevant stratification variables which can be difficult.
2. Is not useful when there are no homogeneous subgroups.
3. Can be expensive to implement.

Poststratification

Stratification is sometimes introduced after the sampling phase in a process called "poststratification". This approach is typically implemented due to a lack of prior knowledge of an appropriate stratifying variable or when the experimenter lacks the necessary information to create a stratifying variable during the sampling phase. Although the method is susceptible to the pitfalls of post hoc approaches, it can provide several benefits in the right situation. Implementation usually follows a simple random sample. In addition to allowing for stratification on an ancillary variable, poststratification can be used to implement weighting, which can improve the precision of a sample's estimates.

Oversampling

Choice-based sampling is one of the stratified sampling strategies. In choice-based sampling, the data are stratified on the target and a sample is taken from each stratum so that the rare target class will be more represented in the sample. The model is then built on this biased sample. The effects of the input variables on the target are often estimated with more precision with the choice-based sample even when a smaller overall sample size is taken, compared to a random sample. The results usually must be adjusted to correct for the oversampling.

Probability-Proportional-to-Size Sampling

In some cases the sample designer has access to an "auxiliary variable" or "size measure", believed to be correlated to the variable of interest, for each element in the population. These data can be used to improve accuracy in sample design. One option is to use the auxiliary variable as a basis for stratification, as discussed above.

Another option is probability proportional to size ('PPS') sampling, in which the selection probability for each element is set to be proportional to its size measure, up to a maximum of 1. In a simple PPS design, these selection probabilities can then be used as the basis for Poisson sampling. However, this has the drawback of variable sample size, and different portions of the population may still be over- or under-represented due to chance variation in selections.

Systematic sampling theory can be used to create a probability proportionate to size sample. This is done by treating each count within the size variable as a single sampling unit. Samples are then identified by selecting at even intervals among these counts within the size variable. This method is sometimes called PPS-sequential or monetary unit sampling in the case of audits or forensic sampling.

Example: Suppose we have six schools with populations of 150, 180, 200, 220, 260, and 490 students respectively (total 1500 students), and we want to use student population as the basis for a PPS sample of size three. To do this, we could allocate the first school numbers 1 to 150, the second school 151 to 330 (= 150 + 180), the third school 331 to 530, and so on to the last school (1011 to 1500). We then generate a random start between 1 and 500 (equal to 1500/3) and count through the school populations by multiples of 500. If our random start was 137, we would select the schools which have been allocated numbers 137, 637, and 1137, i.e. the first, fourth, and sixth schools.

The PPS approach can improve accuracy for a given sample size by concentrating sample on large elements that have the greatest impact on population estimates. PPS sampling is commonly used for surveys of businesses, where element size varies greatly and auxiliary information is often available—for instance, a survey attempting to measure the number of guest-nights spent in hotels might use each hotel's number of rooms as an auxiliary variable. In some cases, an older measurement of the variable of interest can be used as an auxiliary variable when attempting to produce more current estimates.

Cluster Sampling

Sometimes it is more cost-effective to select respondents in groups ('clusters'). Sampling is often clustered by geography, or by time periods. (Nearly all samples are in some sense 'clustered' in time - although this is rarely taken into account in the analysis.) For instance, if surveying households within a city, we might choose to select 100 city blocks and then interview every household within the selected blocks.

Clustering can reduce travel and administrative costs. In the example above, an interviewer can make a single trip to visit several households in one block, rather than having to drive to a different block for each household.

It also means that one does not need a sampling frame listing all elements in the target population. Instead, clusters can be chosen from a cluster-level frame, with an element-level frame created only for the selected clusters. In the example above, the sample only requires a block-level city map for initial selections, and then a household-level map of the 100 selected blocks, rather than a household-level map of the whole city.

Cluster sampling generally increases the variability of sample estimates above that of simple random sampling, depending on how the clusters differ between themselves, as compared with the within-cluster variation. For this reason, cluster sampling requires a larger sample than SRS to achieve the same level of accuracy - but cost savings from clustering might still make this a cheaper option.

Cluster sampling is commonly implemented as multistage sampling. This is a complex form of cluster sampling in which two or more levels of units are embedded one in the other. The first stage consists of constructing the clusters that will be used to sample from. In the second stage, a sample of primary units is randomly selected from each cluster (rather than using all units contained in all selected clusters). In following stages, in each of those selected clusters, additional samples of units are selected, and so on. All ultimate units (individuals, for instance) selected at the last step of this procedure are then surveyed. This technique, thus, is essentially the process of taking random subsamples of preceding random samples.

Multistage sampling can substantially reduce sampling costs, where the complete population list would need to be constructed (before other sampling methods could be applied). By eliminating the work involved in describing clusters that are not selected, multistage sampling can reduce the large costs associated with traditional cluster sampling. However, each sample may not be a full representative of the whole population.

Quota Sampling

In quota sampling, the population is first segmented into mutually exclusive sub-groups, just as in stratified sampling. Then judgement is used to select the subjects or units from each segment based on a specified proportion. For example, an interviewer may be told to sample 200 females and 300 males between the age of 45 and 60.

It is this second step which makes the technique one of non-probability sampling. In quota sampling the selection of the sample is non-random. For example interviewers might be tempted to interview those who look most helpful. The problem is that these samples may be biased because not everyone gets a chance of selection. This random element is its greatest weakness and quota versus probability has been a matter of controversy for several years.

Minimax Sampling

In imbalanced datasets, where the sampling ratio does not follow the population statistics, one can resample the dataset in a conservative manner called minimax sampling. The minimax sampling has its origin in Anderson minimax ratio whose value is proved to be 0.5: in a binary classification, the class-sample sizes should be chosen equally. This ratio can be proved to be minimax ratio only under the assumption of LDA classifier with Gaussian distributions. The notion of minimax sampling is recently developed for a general class of classification rules, called class-wise smart classifiers. In this case, the sampling ratio of classes is selected so that the worst case classifier error over all the possible population statistics for class prior probabilities, would be the best.

Accidental Sampling

Accidental sampling (sometimes known as grab, convenience or opportunity sampling) is a type of nonprobability sampling which involves the sample being drawn from that part of the population which is close to hand. That is, a population is selected because it is readily available and convenient. It may be through meeting the person or including a person in the sample when one meets them or chosen by finding them through technological means such as the internet or through phone. T

he researcher using such a sample cannot scientifically make generalizations about the total population from this sample because it would not be representative enough. For example, if the interviewer were to conduct such a survey at a shopping centre early in the morning on a given day, the people that he/she could interview would be limited to those given there at that given time, which would not represent the views of other members of society in such an area, if the survey were to be conducted at different times of day and several times per week. This type of sampling is most useful for pilot testing. Several important considerations for researchers using convenience samples include:

1. Are there controls within the research design or experiment which can serve to lessen the impact of a non-random convenience sample, thereby ensuring the results will be more representative of the population?
2. Is there good reason to believe that a particular convenience sample would or should respond or behave differently than a random sample from the same population?
3. Is the question being asked by the research one that can adequately be answered using a convenience sample?

In social science research, snowball sampling is a similar technique, where existing study subjects are used to recruit more subjects into the sample. Some variants of snowball sampling, such as respondent driven sampling, allow calculation of selection probabilities and are probability sampling methods under certain conditions.

Line-Intercept Sampling

Line-intercept sampling is a method of sampling elements in a region whereby an element is sampled if a chosen line segment, called a "transect", intersects the element.

Panel Sampling

Panel sampling is the method of first selecting a group of participants through a random sampling method and then asking that group for (potentially the same) information several times over a period of time. Therefore, each participant is interviewed at two or more time points; each period of data collection is called a "wave". The method was developed by sociologist Paul Lazarsfeld in 1938 as a means of studying political campaigns.

This longitudinal sampling-method allows estimates of changes in the population, for example with regard to chronic illness to job stress to weekly food expenditures. Panel sampling can also be used to inform researchers about within-person health changes due to age or to help explain changes in continuous dependent variables such as spousal interaction. There have been several proposed methods of analyzing panel data, including MANOVA, growth curves, and structural equation modelling with lagged effects.

Replacement of Selected Units

Sampling schemes may be *without replacement* ('WOR'—no element can be selected more than once in the same sample) or *with replacement* ('WR'—an element may appear multiple times in the one sample). For example, if we catch fish, measure them, and immediately return them to the water before continuing with the sample, this is a WR design, because we might end up catching and measuring the same fish more than once. However, if we do not return the fish to the water (e.g., if we eat the fish), this becomes a WOR design.

Sample Size

Formulas, tables, and power function charts are well known approaches to determine sample size.

Steps for Using Sample Size Tables

1. Postulate the effect size of interest, α, and β.
2. Check sample size table

1. Select the table corresponding to the selected α
2. Locate the row corresponding to the desired power
3. Locate the column corresponding to the estimated effect size.

4. The intersection of the column and row is the minimum sample size required.

Sampling and Data Collection

Good data collection involves:

- Following the defined sampling process
- Keeping the data in time order
- Noting comments and other contextual events
- Recording non-responses

Errors in Sample Surveys

Survey results are typically subject to some error. Total errors can be classified into sampling errors and non-sampling errors. The term "error" here includes systematic biases as well as random errors.

Sampling Errors and Biases

Sampling errors and biases are induced by the sample design. They include:

1. *Selection bias:* When the true selection probabilities differ from those assumed in calculating the results.
2. *Random sampling error:* Random variation in the results due to the elements in the sample being selected at random.

Non-Sampling Error

Non-sampling errors are other errors which can impact the final survey estimates, caused by problems in data collection, processing, or sample design. They include:

1. Overcoverage: Inclusion of data from outside of the population.
2. Undercoverage: Sampling frame does not include elements in the population.
3. Measurement error: e.g. when respondents misunderstand a question, or find it difficult to answer.
4. Processing error: Mistakes in data coding.
5. Non-response: Failure to obtain complete data from all selected individuals.

After sampling, a review should be held of the exact process followed in sampling, rather than that intended, in order to study any effects that any divergences might have on subsequent analysis. A particular problem is that of *non-response*.

Two major types of nonresponse exist: unit nonresponse (referring to lack of completion of any part of the survey) and item nonresponse (submission or participation in survey but failing to complete one or more components/questions of the survey). In survey sampling, many of the individuals identified as part of the sample may be unwilling to participate, not have the time to participate (opportunity cost), or survey administrators may not have been able to contact them.

In this case, there is a risk of differences, between respondents and nonrespondents, leading to biased estimates of population parameters. This is often addressed by improving survey design, offering incentives, and conducting follow-up studies which make a repeated attempt to contact the unresponsive and to characterize their similarities and differences with the rest of the frame. The effects can also be mitigated by weighting the data when population benchmarks are available or by imputing data based on answers to other questions.

Nonresponse is particularly a problem in internet sampling. Reasons for this problem include improperly designed surveys, over-surveying (or survey fatigue), and the fact that potential participants hold multiple e-mail addresses, which they don't use anymore or don't check regularly.

Survey Weights

In many situations the sample fraction may be varied by stratum and data will have to be weighted to correctly represent the population. Thus for example, a simple random sample of individuals in the United Kingdom might include some in remote Scottish islands who would be inordinately expensive to sample. A cheaper method would be to use a stratified sample with urban and rural strata. The rural sample could be under-represented in the sample, but weighted up appropriately in the analysis to compensate.

More generally, data should usually be weighted if the sample design does not give each individual an equal chance of being selected. For instance, when households have equal selection probabilities but one person is interviewed from within each household, this gives people from large households a smaller chance of being interviewed. This can be accounted for using survey weights. Similarly, households with more than one telephone line have a greater chance of being selected in a random digit dialing sample, and weights can adjust for this.

Weights can also serve other purposes, such as helping to correct for non-response.

Methods of Producing Random Samples

- Random number table
- Mathematical algorithms for pseudo-random number generators
- Physical randomization devices such as coins, playing cards or sophisticated devices such as ERNIE

History

Random sampling by using lots is an old idea, mentioned several times in the Bible. In 1786 Pierre Simon Laplace estimated the population of France by using a sample, along with ratio estimator. He also computed probabilistic estimates of the error. These were not expressed as modern confidence intervals but as the sample size that would be needed to achieve a particular upper bound on the sampling error with probability 1000/1001. His estimates used Bayes' theorem with a uniform prior probability and assumed that his sample was random. Alexander Ivanovich Chuprov introduced sample surveys to Imperial Russia in the 1870s.

In the USA the 1936 *Literary Digest* prediction of a Republican win in the presidential election went badly awry, due to severe bias. More than two million people responded to the study with their names obtained through magazine subscription lists and telephone directories. It was not appreciated that these lists were heavily biased towards Republicans and the resulting sample, though very large, was deeply flawed.

Sampling Error

In statistics, sampling error is incurred when the statistical characteristics of a population are estimated from a subset, or sample, of that population. Since the sample does not include all members of the population, statistics on the sample, such as means and quantiles, generally differ from parameters on the entire population. For example, if one measures the height of a thousand individuals from a country of one million, the average height of the thousand is typically not the same as the average height of all one million people in the country. Since sampling is typically done to determine the characteristics of a whole population, the difference between the sample and population values is considered a *sampling error*. Exact measurement of sampling error is generally not feasible since the true population values are unknown; however, sampling error can often be estimated by probabilistic modelling of the sample.

Description

Random Sampling

In statistics, sampling error is the error caused by observing a sample instead of the whole population. The sampling error can be found by subtracting the value of a parameter from the value of a statistic. In nursing research, a sampling error is the difference between a sample statistic used to estimate a population parameter and the actual but unknown value of the parameter (Bunns & Grove, 2009). An *estimate of a quantity* of interest, such as an average or percentage, will generally be subject to sample-to-sample variation. These variations in the possible sample values of a statistic can theoretically be expressed as sampling errors, although in practice the exact sampling error is typically unknown. Sampling error also refers more broadly to this phenomenon of random sampling variation.

Random sampling, and its derived terms such as sampling error, imply specific procedures for gathering and analyzing data that are rigorously applied as a method for arriving at results considered representative of a given population as a whole. Despite a common misunderstanding, "random" does not mean the same thing as "chance" as this idea is often used in describing situations of uncertainty, nor is it the same as projections based on an assessed probability or frequency. Sampling always refers to a procedure of gathering data from a small aggregation of individuals that is purportedly representative of a larger grouping which must in principle be capable of being measured as a totality. Random sampling is used precisely to ensure a truly representative sample from which to draw conclusions, in which the same results would be arrived at if one had included the entirety of the population instead. Random sampling (and sampling error) can only be used to gather information about a single defined point in time. If additional data is gathered (other things remaining constant) then comparison across time periods may be possible. However, this comparison is distinct from any sampling itself. As a method for gathering data within the field of statistics, random sampling is recognised as clearly distinct from the causal process that one is trying to measure. The conducting of research itself may lead to certain outcomes affecting the researched group, but this effect is not what is called sampling error. Sampling error always refers to the recognised limitations of any supposedly representative sample population in reflecting the larger totality, and the error refers only to the discrepancy that may result from judging the whole on the basis of a much smaller number. This is only an "error" in the sense that it would automatically be corrected if the totality were itself assessed. The term has no real meaning outside of statistics.

According to a differing view, a potential example of a sampling error in evolution is genetic drift; a change is a population's allele frequencies due to chance. For example the bottleneck effect; when natural disasters dramatically reduce the size of a population resulting in a small population that may or may not fairly represent the original population. What may make the bottleneck effect a sampling error is that certain alleles, due to natural disaster, are more common while others may disappear completely, making it a potential sampling error. Another example of genetic drift that is a potential sampling error is the founder effect. The founder effect is when a few individuals from a larger population settle a new isolated area. In this instance, there are only a few individuals with little gene variety, making it a potential sampling error.

The likely size of the sampling error can generally be controlled by taking a large enough random sample from the population, although the cost of doing this may be prohibitive. If the observations are collected from a random sample, statistical theory provides probabilistic estimates of the likely size of the sampling error for a particular statistic or estimator. These are often expressed in terms of its standard error.

Bias Problems

Sampling bias is a possible source of sampling errors. It leads to sampling errors which either have a prevalence to be positive or negative. Such errors can be considered to be systematic errors.

Non-Sampling Error

Sampling error can be contrasted with non-sampling error. Non-sampling error is a catch-all term for the deviations from the true value that are not a function of the sample chosen, including various systematic errors and any random errors that are not due to sampling. Non-sampling errors are much harder to quantify than sampling error.

Chapter 6

Statistical Hypothesis Testing

A statistical hypothesis test is a method of statistical inference using data from a scientific study. In statistics, a result is called statistically significant if it has been predicted as unlikely to have occurred by chance alone, according to a pre-determined threshold probability, the significance level. The phrase "test of significance" was coined by statistician Ronald Fisher. These tests are used in determining what outcomes of a study would lead to a rejection of the null hypothesis for a pre-specified level of significance; this can help to decide whether results contain enough information to cast doubt on conventional wisdom, given that conventional wisdom has been used to establish the null hypothesis.

The *critical region* of a hypothesis test is the set of all outcomes which cause the null hypothesis to be rejected in favour of the alternative hypothesis. Statistical hypothesis testing is sometimes called confirmatory data analysis, in contrast to exploratory data analysis, which may not have pre-specified hypotheses. In the Neyman-Pearson framework, the process of distinguishing between the null & alternative hypotheses is aided by identifying two conceptual types of errors (type 1 & type 2), and by specifying parametric limits on e.g. how much type 1 error will be permitted.

Variations and Sub-Classes

Statistical hypothesis testing is a key technique of both Frequentist inference and Bayesian inference although they have notable differences. Statistical hypothesis tests define a procedure that controls (fixes) the probability of incorrectly *deciding* that a default position (null hypothesis) is incorrect based on how likely it would be for a set of observations to occur if the null hypothesis were true. Note that this probability of making an incorrect decision is not the probability that the null hypothesis is true, nor whether any specific alternative hypothesis is true. This contrasts with other possible techniques of decision theory in which the null and alternative hypothesis are treated on a more equal basis. One naive Bayesian approach to hypothesis testing is to base decisions on the posterior probability, but this fails when comparing point and continuous hypotheses. Other approaches to decision making, such as Bayesian decision theory, attempt to balance the consequences of incorrect decisions across all possibilities, rather than concentrating on a single null hypothesis. A number of other approaches to reaching a decision based on data are available via decision theory and optimal decisions, some of which have desirable properties, yet hypothesis testing is a dominant approach to data analysis in many fields of science. Extensions to the theory of hypothesis testing include the study of the power of tests, the probability of correctly rejecting the null hypothesis given that it is false. Such considerations can be used for the purpose of sample size determination prior to the collection of data.

The Testing Process

In the statistics literature, statistical hypothesis testing plays a fundamental role. The usual line of reasoning is as follows:

1. There is an initial research hypothesis of which the truth is unknown.
2. The first step is to state the relevant null and alternative hypotheses. This is important as mis-stating the hypotheses will muddy the rest of the process.
3. The second step is to consider the statistical assumptions being made about the sample in doing the test; for example, assumptions about the statistical independence or about the form of the distributions of the observations. This is equally important as invalid assumptions will mean that the results of the test are invalid.
4. Decide which test is appropriate, and state the relevant test statistic T.
5. Derive the distribution of the test statistic under the null hypothesis from the assumptions. In standard cases this will be a well-known result. For example the test statistic might follow a Student's t distribution or a normal distribution.

6. Select a significance level (α), a probability threshold below which the null hypothesis will be rejected. Common values are 5% and 1%.
7. The distribution of the test statistic under the null hypothesis partitions the possible values of T into those for which the null hypothesis is rejected, the so-called critical region, and those for which it is not. The probability of the critical region is α.
8. Compute from the observations the observed value t_{obs} of the test statistic T.
9. Decide to either reject the null hypothesis in favour of the alternative or not reject it. The decision rule is to reject the null hypothesis H_0 if the observed value t_{obs} is in the critical region, and to accept or "fail to reject" the hypothesis otherwise.

An Alternative Process is Commonly Used

1. Compute from the observations the observed value t_{obs} of the test statistic T.
2. Calculate the p-value. This is the probability, under the null hypothesis, of sampling a test statistic at least as extreme as that which was observed.
3. Reject the null hypothesis, in favour of the alternative hypothesis, if and only if the p-value is less than the significance level (the selected probability) threshold.

The two processes are equivalent. The former process was advantageous in the past when only tables of test statistics at common probability thresholds were available. It allowed a decision to be made without the calculation of a probability. It was adequate for classwork and for operational use, but it was deficient for reporting results.

The latter process relied on extensive tables or on computational support not always available. The explicit calculation of a probability is useful for reporting. The calculations are now trivially performed with appropriate software.

The difference in the two processes applied to the Radioactive suitcase example (below):

- "The Geiger-counter reading is 10. The limit is 9. Check the suitcase."
- "The Geiger-counter reading is high; 97% of safe suitcases have lower readings. The limit is 95%. Check the suitcase."

The former report is adequate, the latter gives a more detailed explanation of the data and the reason why the suitcase is being checked.

It is important to note the philosophical difference between accepting the null hypothesis and simply failing to reject it. The "fail to reject" terminology highlights the fact that the null hypothesis is assumed to be true from the start of the test; if there is a lack of evidence against it, it simply continues to be assumed true.

The phrase "accept the null hypothesis" may suggest it has been proved simply because it has not been disproved, a logical fallacy known as the argument from ignorance. Unless a test with particularly high power is used, the idea of "accepting" the null hypothesis may be dangerous. Nonetheless the terminology is prevalent throughout statistics, where its meaning is well understood.

The processes described here are perfectly adequate for computation. They seriously neglect the design of experiments considerations.

It is particularly critical that appropriate sample sizes be estimated before conducting the experiment.

Interpretation

If the p-value is less than the required significance level (equivalently, if the observed test statistic is in the critical region), then we say the null hypothesis is rejected at the given level of significance. Rejection of the null hypothesis is a conclusion. This is like a "guilty" verdict in a criminal trial – the evidence is sufficient to reject innocence, thus proving guilt. We might accept the alternative hypothesis (and the research hypothesis).

If the p-value is not less than the required significance level (equivalently, if the observed test statistic is outside the critical region), then the test has no result. The evidence is insufficient to support a conclusion. (This is like a jury that fails to reach a verdict.) The researcher typically gives extra consideration to those cases where the p-value is close to the significance level.

In the Lady tasting tea example (below), Fisher required the Lady to properly categorize all of the cups of tea to justify the conclusion that the result was unlikely to result from chance. He defined the critical region as that case alone. The region was defined by a probability (that the null hypothesis was correct) of less than 5%.

Whether rejection of the null hypothesis truly justifies acceptance of the research hypothesis depends on the structure of the hypotheses. Rejecting the hypothesis that a large paw print originated from a bear does not immediately prove the existence of Bigfoot. Hypothesis testing emphasizes the rejection which is based on a probability rather than the acceptance which requires extra steps of logic.

"The probability of rejecting the null hypothesis is a function of five factors: whether the test is one- or two tailed, the level of significance, the standard deviation, the amount of deviation from the null hypothesis, and the number of observations." These factors are a source of criticism; factors under the control of the experimenter/analyst give the results an appearance of subjectivity.

Use and Importance

Statistics are helpful in analyzing most collections of data. This is equally true of hypothesis testing which can justify conclusions even when no scientific theory exists. In the Lady tasting tea example, it was "obvious" that no difference existed between (milk poured into tea) and (tea poured into milk). The data contradicted the "obvious".

Real world applications of hypothesis testing include:

- Testing whether more men than women suffer from nightmares
- Establishing authorship of documents
- Evaluating the effect of the full moon on behaviour
- Determining the range at which a bat can detect an insect by echo
- Deciding whether hospital carpeting results in more infections
- Selecting the best means to stop smoking
- Checking whether bumper stickers reflect car owner behaviour
- Testing the claims of handwriting analysts

Statistical hypothesis testing plays an important role in the whole of statistics and in statistical inference. For example, Lehmann (1992) in a review of the fundamental paper by Neyman and Pearson (1933) says: "Nevertheless, despite their shortcomings, the new paradigm formulated in the 1933 paper, and the many developments carried out within its framework continue to play a central role in both the theory and practice of statistics and can be expected to do so in the foreseeable future".

Significance testing has been the favoured statistical tool in some experimental social sciences (over 90% of articles in the Journal of Applied Psychology during the early 1990s). Other fields have favoured the estimation of parameters (e.g., effect size). Significance testing is used as a substitute for the traditional comparison of predicted value and experimental result at the core of the scientific method. When theory is only capable of predicting the sign of a relationship, a directional (one-sided) hypothesis test can be configured so that only a statistically significant result supports theory. This form of theory appraisal is the most heavily criticized application of hypothesis testing.

Cautions

"If the government required statistical procedures to carry warning labels like those on drugs, most inference methods would have long labels indeed." This caution applies to hypothesis tests and alternatives to them.

The successful hypothesis test is associated with a probability and a type-I error rate. The conclusion *might* be wrong.

The conclusion of the test is only as solid as the sample upon which it is based. The design of the experiment is critical. A number of unexpected effects have been observed including:

- The Clever Hans effect. A horse appeared to be capable of doing simple arithmetic.
- The Hawthorne effect. Industrial workers were more productive in better illumination, and most productive in worse.
- The Placebo effect. Pills with no medically active ingredients were remarkably effective.

A statistical analysis of misleading data produces misleading conclusions. The issue of data quality can be more subtle. In forecasting for example, there is no agreement on a measure of forecast accuracy. In the absence of a consensus measurement, no decision based on measurements will be without controversy.

The book *How to Lie with Statistics* is the most popular book on statistics ever published. It does not much consider hypothesis testing, but its cautions are applicable, including: Many claims are made on the basis of samples too small to convince. If a report does not mention sample size, be doubtful.

Hypothesis testing acts as a filter of statistical conclusions; only those results meeting a probability threshold are publishable. Economics also acts as a publication filter; only those results favourable to the author and funding source may be submitted for publication. The impact of filtering on publication is termed publication bias. A related problem is that of multiple testing (sometimes linked to data mining), in which a variety of tests for a variety of possible effects are applied to a single data set and only those yielding a significant result are reported. These are often dealt with by using multiplicity correction procedures that control the family wise error rate (FWER) or the false discovery rate (FDR).

Those making critical decisions based on the results of a hypothesis test are prudent to look at the details rather than the conclusion alone. In the physical sciences most results are fully accepted only when independently confirmed. The general advice concerning statistics is, "Figures never lie, but liars figure" (anonymous).

Example

Lady Tasting Tea

In a famous example of hypothesis testing, known as the *Lady tasting tea* example, a female colleague of Fisher claimed to be able to tell whether the tea or the milk was added first to a cup. Fisher proposed to give her eight cups, four of each variety, in random order. One could then ask what the probability was for her getting the number she got correct, but just by chance.

The null hypothesis was that the Lady had no such ability. The test statistic was a simple count of the number of successes in selecting the 4 cups. The critical region was the single case of 4 successes of 4 possible based on a conventional probability criterion ($< 5\%$; 1 of 70 $\approx 1.4\%$). Fisher asserted that no alternative hypothesis was (ever) required. The lady correctly identified every cup, which would be considered a statistically significant result.

Analogy – Courtroom Trial

A statistical test procedure is comparable to a criminal trial; a defendant is considered not guilty as long as his or her guilt is not proven. The prosecutor tries to prove the guilt of the defendant. Only when there is enough charging evidence the defendant is convicted.

In the start of the procedure, there are two hypotheses H_0: "the defendant is not guilty", and H_1: "the defendant is guilty". The first one is called *null hypothesis*, and is for the time being accepted. The second one is called *alternative (hypothesis)*. It is the hypothesis one hopes to support.

The hypothesis of innocence is only rejected when an error is very unlikely, because one doesn't want to convict an innocent defendant.

Such an error is called *error of the first kind* (i.e., the conviction of an innocent person), and the occurrence of this error is controlled to be rare. As a consequence of this asymmetric behaviour, the *error of the second kind* (acquitting a person who committed the crime), is often rather large.

	H_0 is true ***Truly not guilty***	***H_1 is true*** ***Truly guilty***
Accept Null Hypothesis Acquittal	Right decision	Wrong decision Type II Error
Reject Null Hypothesis Conviction	Wrong decision Type I Error	Right decision

A criminal trial can be regarded as either or both of two decision processes: guilty vs not guilty or evidence vs a threshold ("beyond a reasonable doubt").

In one view, the defendant is judged; in the other view the performance of the prosecution (which bears the burden of proof) is judged. A hypothesis test can be regarded as either a judgment of a hypothesis or as a judgment of evidence.

Example 1 – Philosopher's Beans

The following example was produced by a philosopher describing scientific methods generations before hypothesis testing was formalized and popularized.

Few beans of this handful are white.

Most beans in this bag are white.

Therefore: Probably, these beans were taken from another bag.

This is an hypothetical inference.

The beans in the bag are the population. The handful are the sample. The null hypothesis is that the sample originated from the population.

The criterion for rejecting the null-hypothesis is the "obvious" difference in appearance (an informal difference in the mean). The interesting result is that consideration of a real population and a real sample produced an imaginary bag. The philosopher was considering logic rather than probability.

To be a real statistical hypothesis test, this example requires the formalities of a probability calculation and a comparison of that probability to a standard.

A simple generalization of the example considers a mixed bag of beans and a handful that contain either very few or very many white beans. The generalization considers both extremes.

It requires more calculations and more comparisons to arrive at a formal answer, but the core philosophy is unchanged; If the composition of the handful is greatly different from that of the bag, then the sample probably originated from another bag.

The original example is termed a one-sided or a one-tailed test while the generalization is termed a two-sided or two-tailed test.

The statement also relies on the inference that the sampling was random. If someone had been picking through the bag to find white beans, then it would explain why the handful had so many white beans, and also explain why the number of white beans in the bag was depleted (although the bag is probably intended to be assumed much larger than one's hand).

Example 2 – Clairvoyant Card Game

A person (the subject) is tested for clairvoyance. He is shown the reverse of a randomly chosen playing card 25 times and asked which of the four suits it belongs to. The number of hits, or correct answers, is called X.

As we try to find evidence of his clairvoyance, for the time being the null hypothesis is that the person is not clairvoyant. The alternative is, of course: the person is (more or less) clairvoyant. If the null hypothesis is valid, the only thing the test person can do is guess. For every card, the probability (relative frequency) of any single suit appearing is 1/4.

If the alternative is valid, the test subject will predict the suit correctly with probability greater than 1/4. We will call the probability of guessing correctly p. The hypotheses, then, are:

- null hypothesis : $H_0 : p = \frac{1}{4}$ (just guessing)

and

- alternative hypothesis :$H_1 : p > \frac{1}{4}$ (true clairvoyant).

When the test subject correctly predicts all 25 cards, we will consider him clairvoyant, and reject the null hypothesis. Thus also with 24 or 23 hits. With only 5 or 6 hits, on the other hand, there is no cause to consider him so. But what about 12 hits, or 17 hits? What is the critical number, c, of hits, at which point we consider the subject to be clairvoyant? How do we determine the critical value c? It is obvious that with the choice c=25 (i.e. we only accept clairvoyance when all cards are predicted correctly) we're more critical than with c=10.

In the first case almost no test subjects will be recognised to be clairvoyant, in the second case, a certain number will pass the test. In practice, one decides how critical one will be.

That is, one decides how often one accepts an error of the first kind – a false positive, or Type I error. With $c = 25$ the probability of such an error is:

$$P(\text{reject } H_0 \mid H_0 \text{ is valid}) = P(X = 25 \mid p = \tfrac{1}{4}) = \left(\tfrac{1}{4}\right)^{25} \approx 10^{-15},$$

and hence, very small. The probability of a false positive is the probability of randomly guessing correctly all 25 times.

Being less critical, with c=10, gives:

$$P(\text{reject } H_0 \mid H_0 \text{ is valid}) = P(X \geq 10 \mid p = \tfrac{1}{4}) = \sum_{k=10}^{25} P(X = k \mid p = \tfrac{1}{4}) \approx 0.07.$$

Thus, $c = 10$ yields a much greater probability of false positive.

Before the test is actually performed, the maximum acceptable probability of a Type I error (α) is determined. Typically, values in the range of 1% to 5% are selected.

(If the maximum acceptable error rate is zero, an infinite number of correct guesses is required.)

Depending on this Type 1 error rate, the critical value c is calculated. For example, if we select an error rate of 1%, c is calculated thus:

$$P(\text{reject } H_0 \mid H_0 \text{ is valid}) = P(X \geq c \mid p = \tfrac{1}{4}) \leq 0.01.$$

From all the numbers c, with this property, we choose the smallest, in order to minimize the probability of a Type II error, a false negative. For the above example, we select: $c = 13$.

Example 3 – Radioactive Suitcase

As an example, consider determining whether a suitcase contains some radioactive material. Placed under a Geiger counter, it produces 10 counts per minute.

The null hypothesis is that no radioactive material is in the suitcase and that all measured counts are due to ambient radioactivity typical of the surrounding air and harmless objects. We can then calculate how likely it is that we would observe 10 counts per minute if the null hypothesis were true. If the null hypothesis predicts (say) on average 9 counts per minute, then according to the Poisson distribution typical for radioactive decay there is about 41% chance of recording 10 or more counts.

Thus we can say that the suitcase is compatible with the null hypothesis (this does not guarantee that there is no radioactive material, just that we don't have enough evidence to suggest there is).

On the other hand, if the null hypothesis predicts 3 counts per minute (for which the Poisson distribution predicts only 0.1% chance of recording 10 or more counts) then the suitcase is not compatible with the null hypothesis, and there are likely other factors responsible to produce the measurements.

The test does not directly assert the presence of radioactive material. A *successful* test asserts that the claim of no radioactive material present is unlikely given the reading (and therefore ...). The double negative (disproving the null hypothesis) of the method is confusing, but using a counter-example to disprove is standard mathematical practice.

The attraction of the method is its practicality. We know (from experience) the expected range of counts with only ambient radioactivity present, so we can say that a measurement is *unusually* large. Statistics just formalizes the intuitive by using numbers instead of adjectives. We probably do not know the characteristics of the radioactive suitcases; We just assume that they produce larger readings.

To slightly formalize intuition: Radioactivity is suspected if the Geiger-count with the suitcase is among or exceeds the greatest (5% or 1%) of the Geiger-counts made with ambient radiation alone. This makes no assumptions about the distribution of counts. Many ambient radiation observations are required to obtain good probability estimates for rare events.

The test described here is more fully the null-hypothesis statistical significance test. The null hypothesis represents what we would believe by default, before seeing any evidence. Statistical significance is a possible finding of the test, declared when the observed sample is unlikely to have occurred by chance if the null hypothesis were true. The name of the test describes its formulation and its possible outcome. One characteristic of the test is its crisp decision: to reject or not reject the null hypothesis.

A calculated value is compared to a threshold, which is determined from the tolerable risk of error.

Definition of Terms

The following definitions are mainly based on the exposition in the book by Lehmann and Romano:

Statistical Hypothesis: A statement about the parameters describing a population (not a sample).

Statistic: A value calculated from a sample, often to summarize the sample for comparison purposes.

Simple Hypothesis: Any hypothesis which specifies the population distribution completely.

Composite Hypothesis: Any hypothesis which does *not* specify the population distribution completely.

Null Hypothesis (H_0): A simple hypothesis associated with a contradiction to a theory one would like to prove.

Alternative Hypothesis (H_1): A hypothesis (often composite) associated with a theory one would like to prove.

Statistical Test: A procedure whose inputs are samples and whose result is a hypothesis.

Region of Acceptance: The set of values of the test statistic for which we fail to reject the null hypothesis.

Region of Rejection / Critical Region: The set of values of the test statistic for which the null hypothesis is rejected.

Critical Value: The threshold value delimiting the regions of acceptance and rejection for the test statistic.

Power of a Test ($1 - \beta$): The test's probability of correctly rejecting the null hypothesis. The complement of the false negative rate, β. Power is termed sensitivity in biostatistics. ("This is a sensitive test. Because the result is negative, we can confidently say that the patient does not have the condition.")

Size / Significance Level of a Test (α): For simple hypotheses, this is the test's probability of *incorrectly* rejecting the null hypothesis. The false positive rate. For composite hypotheses this is the upper bound of the probability of rejecting the null hypothesis over all cases covered by the null hypothesis. The complement of the false positive rate, (1 " α), is termed specificity in biostatistics. ("This is a specific test. Because the result is positive, we can confidently say that the patient has the condition.").

p-value: The probability, assuming the null hypothesis is true, of observing a result at least as extreme as the test statistic.

Statistical Significance Test : A predecessor to the statistical hypothesis test. An experimental result was said to be statistically significant if a sample was sufficiently inconsistent with the (null) hypothesis. This was variously considered common sense, a pragmatic heuristic for identifying meaningful experimental results, a convention establishing a threshold of statistical evidence or a method for drawing conclusions from data.

The statistical hypothesis test added mathematical rigor and philosophical consistency to the concept by making the alternative hypothesis explicit. The term is loosely used to describe the modern version which is now part of statistical hypothesis testing.

Conservative Test: A test is conservative if, when constructed for a given nominal significance level, the true probability of *incorrectly* rejecting the null hypothesis is never greater than the nominal level.

Exact Test: A test in which the significance level or critical value can be computed exactly, i.e., without any approximation. In some contexts this term is restricted to tests applied to categorical data and to permutation tests, in which computations are carried out by complete enumeration of all possible outcomes and their probabilities.

A statistical hypothesis test compares a test statistic (z or t for examples) to a threshold. The test statistic (the formula found in the table below) is based on optimality. For a fixed level of Type I error rate, use of these statistics minimizes Type II error rates (equivalent to maximizing power). The following terms describe tests in terms of such optimality:

Most Powerful Test: For a given *size* or *significance level*, the test with the greatest power (probability of rejection) for a given value of the parameter(s) being tested, contained in the alternative hypothesis.

Uniformly Most Powerful Test (UMP): A test with the greatest *power* for all values of the parameter(s) being tested, contained in the alternative hypothesis.

Common Test Statistics

One-sample tests are appropriate when a sample is being compared to the population from a hypothesis. The population characteristics are known from theory or are calculated from the population.

Two-sample tests are appropriate for comparing two samples, typically experimental and control samples from a scientifically controlled experiment.

Paired tests are appropriate for comparing two samples where it is impossible to control important variables. Rather than comparing two sets, members are paired between samples so the difference between the members becomes the sample. Typically the mean of the differences is then compared to zero.

The common example scenario for when a paired difference test is appropriate is when a single set of test subjects has something applied to them and the test is intended to check for an effect.

Z-tests are appropriate for comparing means under stringent conditions regarding normality and a known standard deviation.

T-tests are appropriate for comparing means under relaxed conditions (less is assumed).

Tests of proportions are analogous to tests of means (the 50% proportion).

Chi-squared tests use the same calculations and the same probability distribution for different applications:

- Chi-squared tests for variance are used to determine whether a normal population has a specified variance. The null hypothesis is that it does.
- Chi-squared tests of independence are used for deciding whether two variables are associated or are independent. The variables are categorical rather than numeric. It can be used to decide whether left-handedness is correlated with libertarian politics (or not). The null hypothesis is that the variables are independent. The numbers used in the calculation are the observed and expected frequencies of occurrence (from contingency tables).
- Chi-squared goodness of fit tests are used to determine the adequacy of curves fit to data. The null hypothesis is that the curve fit is adequate. It is common to determine curve shapes to minimize the mean square error, so it is appropriate that the goodness-of-fit calculation sums the squared errors.

F-tests (analysis of variance, ANOVA) are commonly used when deciding whether groupings of data by category are meaningful. If the variance of test scores of the left-handed in a class is much smaller than the variance of the whole class, then it may be useful to study lefties as a group.

The null hypothesis is that two variances are the same – so the proposed grouping is not meaningful.

In the table below, the symbols used are defined at the bottom of the table. Many other tests can be found in other articles. Proofs exist that the test statistics are appropriate.

Name	*Formula*	*Assumptions or notes*
One-sample z-test	$z = \frac{\overline{x} - \mu_0}{\sigma}\sqrt{n}$	(Normal population or $n > 30$) and σ known. (z is the distance from the mean in relation to the standard deviation of the mean). For non-normal distributions it is possible to calculate a minimum proportion of a population that falls within k standard deviations for any k (see: *Chebyshev's inequality*).
Two-sample z-test	$z = \frac{(\overline{x}_1 - \overline{x}_2) - d_0}{\sqrt{\frac{\sigma_1^2}{n_1} + \frac{\sigma_2^2}{n_2}}}$	Normal population and independent observations and σ_1 and σ_2 are known
One-sample t-test	$t = \frac{\overline{x} - \mu_0}{(s / \sqrt{n})}$, $df = n - 1$	(Normal population or $n > 30$) and σ unknown
Paired t-test	$t = \frac{\overline{d} - d_0}{(s_d / \sqrt{n})}$, $df = n - 1$	(Normal population of differences or $n > 30$) and σ unknown or small sample size $n < 30$
Two-sample pooled t-test, equal variances	$t = \frac{(\overline{x}_1 - \overline{x}_2) - d_0}{s_p\sqrt{\frac{1}{n_1} + \frac{1}{n_2}}}$, $s_p^2 = \frac{(n_1 - 1)s_1^2 + (n_2 - 1)s_2^2}{n_1 + n_2 - 2}$, $df = n_1 + n_2 - 2$	(Normal populations or $n_1 + n_2 > 40$) and independent observations and $\sigma_1 = \sigma_2$ unknown
Two sample unpooled t test, unequal variances (Welch's t test)	$t = \frac{(\overline{x}_1 - \overline{x}_2) - d_0}{\sqrt{\frac{s_1^2}{n_1} + \frac{s_2^2}{n_2}}}$, $df = \frac{\left(\frac{s_1^2}{n_1} + \frac{s_2^2}{n_2}\right)^2}{\frac{\left(\frac{s_1^2}{n_1}\right)^2}{n_1 - 1} + \frac{\left(\frac{s_2^2}{n_2}\right)^2}{n_2 - 1}}$	(Normal populations or $n_1 + n_2 > 40$) and independent observations and σ_1 ? σ_2 both unknown
One-proportion z-test	$z = \frac{\hat{p} - p_0}{\sqrt{p_0(1 - p_0)}}\sqrt{n}$	$n\,p_0 > 10$ and $n\,(1 - p_0) > 10$ and it is a SRS (Simple Random Sample), see notes.
Two-proportion z-test, pooled for $H_0 : p_1 = p_2$	$z = \frac{(\hat{p}_1 - \hat{p}_2)}{\sqrt{\hat{p}(1 - \hat{p})(\frac{1}{n_1} + \frac{1}{n_2})}}$, $\hat{p} = \frac{x_1 + x_2}{n_1 + n_2}$	$n_1 p_1 > 5$ and $n_1(1 - p_1) > 5$ and $n_2 p_2 > 5$ and $n_2(1 - p_2) > 5$ and independent observations, see notes.
Two-proportion z-test, unpooled for $\lvert d_0 \rvert > 0$	$z = \frac{(\hat{p}_1 - \hat{p}_2) - d_0}{\sqrt{\frac{\hat{p}_1(1 - \hat{p}_1)}{n_1} + \frac{\hat{p}_2(1 - \hat{p}_2)}{n_2}}}$	$n_1 p_1 > 5$ and $n_1(1 - p_1) > 5$ and $n_2 p_2 > 5$ and $n_2(1 - p_2) > 5$ and independent observations, see notes.
Chi-squared test for variance	$\chi^2 = (n - 1)\frac{s^2}{\sigma_0^2}$	Normal population
Chi-squared test for goodness of fit	$\chi^2 = \sum^k \frac{(\text{observed} - \text{expected})^2}{\text{expected}}$	$df = k - 1 -$ *# parameters estimated*, and one of these must hold. • All expected counts are at least 5. • All expected counts are > 1 and no more than 20% of expected counts are less than 5
Two-sample F test for equality of variances	$F = \frac{s_1^2}{s_2^2}$	Normal populations Arrange so $s_1^2 \geq s_2^2$ and reject H_0 for $F > F(\alpha / 2, n_1 - 1, n_2 - 1)$
Regression t-test of $H_0 : R^2 = 0.$	$t = \sqrt{\frac{R^2(n - k - 1^*)}{1 - R^2}}$	Reject H_0 for $t > t(\alpha / 2, n - k - 1^*)$ *Subtract 1 for intercept; k terms contain independent variables.

In general, the subscript 0 indicates a value taken from the null hypothesis, H_0, which should be used as much as possible in constructing its test statistic. ... *Definitions of other symbols:*

- α, the probability of Type I error (rejecting a null hypothesis when it is in fact true)
- n= sample size
- $n1$= sample 1 size
- $n2$= sample 2 size
- x= sample mean
- μ_0 = hypothesized population mean
- μ_1 = population 1 mean
- μ_2 = population 2 mean
- σ = population standard deviation
- σ^2 = population variance
- s= sample standard deviation
- $\sum^{k}$ = sum (of k numbers)
- s^2 = sample variance
- s_1 = sample 1 standard deviation
- s_2 = sample 2 standard deviation
- t= t statistic
- df = degrees of freedom
- $\bar{d}$ = sample mean of differences
- d_0 = hypothesized population mean difference
- s_d = standard deviation of differences
- χ^2 = Chi-squared statistic
- $\hat{p} = x/n$ = sample proportion, unless specified otherwise
- p_0 = hypothesized population proportion
- p_1 = proportion 1
- p_2 = proportion 2
- d_p = hypothesized difference in proportion
- $\min\{n_1, n_2\}$ = minimum of n_1 and n_2
- $x_1 = n_1 p_1$

- $x_2 = n_2 p_2$
- F = F statistic

Origins and Early Controversy

Significance testing is largely the product of Karl Pearson (p-value, Pearson's chi-squared test), William Sealy Gosset (Student's t-distribution), and Ronald Fisher ("null hypothesis", analysis of variance, "significance test"), while hypothesis testing was developed by Jerzy Neyman and Egon Pearson (son of Karl). Ronald Fisher, mathematician and biologist described by Richard Dawkins as "the greatest biologist since Darwin", began his life in statistics as a Bayesian (Zabell 1992), but Fisher soon grew disenchanted with the subjectivity involved (namely use of the principle of indifference when determining prior probabilities), and sought to provide a more "objective" approach to inductive inference.

Fisher was an agricultural statistician who emphasized rigorous experimental design and methods to extract a result from few samples assuming Gaussian distributions. Neyman (who teamed with the younger Pearson) emphasized mathematical rigor and methods to obtain more results from many samples and a wider range of distributions. Modern hypothesis testing is an inconsistent hybrid of the Fisher vs Neyman/Pearson formulation, methods and terminology developed in the early 20th century. While hypothesis testing was popularized early in the 20th century, evidence of its use can be found much earlier. In the 1770s Laplace considered the statistics of almost half a million births. The statistics showed an excess of boys compared to girls. He concluded by calculation of a p-value that the excess was a real, but unexplained, effect.

Fisher popularized the "significance test". He required a null-hypothesis (corresponding to a population frequency distribution) and a sample. His (now familiar) calculations determined whether to reject the null-hypothesis or not. Significance testing did not utilise an alternative hypothesis so there was no concept of a Type II error.

The p-value was devised as an informal, but objective, index meant to help a researcher determine (based on other knowledge) whether to modify future experiments or strengthen one's faith in the null hypothesis. Hypothesis testing (and Type I/II errors) was devised by Neyman and Pearson as a more objective alternative to Fisher's p-value, also meant to determine researcher behaviour, but without requiring any inductive inference by the researcher.

Neyman & Pearson considered a different problem (which they called "hypothesis testing"). They initially considered two simple hypotheses (both with frequency distributions). They calculated two probabilities and typically selected the hypothesis associated with the higher probability (the hypothesis more likely to have generated the sample). Their method always selected a hypothesis. It also allowed the calculation of both types of error probabilities.

Fisher and Neyman/Pearson clashed bitterly. Neyman/Pearson considered their formulation to be an improved generalization of significance testing. (The defining paper was abstract. Mathematicians have generalized and refined the theory for decades.) Fisher thought that it was not applicable to scientific research because often, during the course of the experiment, it is discovered that the initial assumptions about the null hypothesis are questionable due to unexpected sources of error. He believed that the use of rigid reject/accept decisions based on models formulated before data is collected was incompatible with this common scenario faced by scientists and attempts to apply this method to scientific research would lead to mass confusion.

The dispute between Fisher and Neyman-Pearson was waged on philosophical grounds, characterized by a philosopher as a dispute over the proper role of models in statistical inference.

Events intervened: Neyman accepted a position in the western hemisphere, breaking his partnership with Pearson and separating disputants (who had occupied the same building) by much of the planetary diameter. World War II provided an intermission in the debate. The dispute between Fisher and Neyman terminated (unresolved after 27 years) with Fisher's death in 1962. Neyman wrote a well-regarded eulogy. Some of Neyman's later publications reported p-values and significance levels.

The modern version of hypothesis testing is a hybrid of the two approaches that resulted from confusion by writers of statistical textbooks (as predicted by Fisher) beginning in the 1940s. (But signal detection, for example, still uses the Neyman/Pearson formulation.) Great conceptual differences and many caveats in addition to those mentioned above were ignored. Neyman and Pearson provided the stronger terminology, the more rigorous mathematics and the more consistent philosophy, but the subject taught today in introductory statistics has more similarities with Fisher's method than theirs. This history explains the inconsistent terminology (example: the null hypothesis is never accepted, but there is a region of acceptance).

Sometime around 1940, in an apparent effort to provide researchers with a "non-controversial" way to have their cake and eat it too, the authors of statistical text books began anonymously combining these two strategies by using the p-value in place of the test statistic (or data) to test against the Neyman-Pearson "significance level". Thus, researchers were encouraged to infer the strength of their data against some null hypothesis using p-values, while also thinking they are retaining the post-data collection objectivity provided by hypothesis testing. It then became customary for the null hypothesis, which was originally some realistic research hypothesis, to be used almost solely as a strawman "nil" hypothesis (one where a treatment has no effect, regardless of the context).

A Comparison between Fisherian, frequentist (Neyman-Pearson)

Fisher's null hypothesis testing	*Neyman–Pearson decision theory*
1. Set up a statistical null hypothesis. The null need not be a nil hypothesis (i.e., zero difference).	1. Set up two statistical hypotheses, H1 and H2, and decide about α, β, and sample size before the experiment, based on subjective cost-benefit considerations. These define a rejection region for each hypothesis.
2. Report the exact level of significance (e.g., p = 0.051 or p = 0.049). Do not use a conventional 5% level, and do not talk about accepting or rejecting hypotheses. If the result is "not significant", draw no conclusions and make no decisions, but suspend judgement until further data is available.	2. If the data falls into the rejection region of H1, accept H2; otherwise accept H1. Note that accepting a hypothesis does not mean that you believe in it, but only that you act as if it were true.
3. Use this procedure only if little is known about the problem at hand, and only to draw provisional conclusions in the context of an attempt to understand the experimental situation.	3. The usefulness of the procedure is limited among others to situations where you have a disjunction of hypotheses (e.g., either μ1 = 8 or μ2 = 10 is true) and where you can make meaningful cost-benefit trade-offs for choosing alpha and beta.

Early Choices of Null Hypothesis

Paul Meehl has argued that the epistemological importance of the choice of null hypothesis has gone largely unacknowledged. When the null hypothesis is predicted by theory, a more precise experiment will be a more severe test of the underlying theory. When the null hypothesis defaults to "no difference" or "no effect", a more precise experiment is a less severe test of the theory that motivated performing the experiment. An examination of the origins of the latter practice may therefore be useful:

1778: Pierre Laplace compares the birthrates of boys and girls in multiple European cities. He states: "it is natural to conclude that these possibilities are very nearly in the same ratio". Thus Laplace's null hypothesis that the birthrates of boys and girls should be equal given "conventional wisdom".

1900: Karl Pearson develops the chi squared test to determine "whether a given form of frequency curve will effectively describe the samples drawn from a given population." Thus the null hypothesis is that a population is described by some distribution predicted by theory. He uses as an example the numbers of five and sixes in the Weldon dice throw data.

1904: Karl Pearson develops the concept of "contingency" in order to determine whether outcomes are independent of a given categorical factor. Here the null hypothesis is by default that two things are unrelated (e.g. scar formation and death rates from smallpox). The null hypothesis in this case is no longer predicted by theory or conventional wisdom, but is instead the principle of indifference that lead Fisher and others to dismiss the use of "inverse probabilities".

Null Hypothesis Statistical Significance Testing vs Hypothesis Testing

An example of Neyman-Pearson hypothesis testing can be made by a change to the radioactive suitcase example. If the "suitcase" is actually a shielded container for the transportation of radioactive material, then a test might be used to select among three hypotheses: no radioactive source present, one present, two (all) present. The test could be required for safety, with actions required in each case. The Neyman-Pearson lemma of hypothesis testing says that a good criterion for the selection of hypotheses is the ratio of their probabilities (a likelihood ratio). A simple method of solution is to select the hypothesis with the highest probability for the Geiger counts observed. The typical result matches intuition: few counts imply no source, many counts imply two sources and intermediate counts imply one source.

Neyman-Pearson theory can accommodate both prior probabilities and the costs of actions resulting from decisions. The former allows each test to consider the results of earlier tests (unlike Fisher's significance tests). The latter allows the consideration of economic issues (for example) as well as probabilities. A likelihood ratio remains a good criterion for selecting among hypotheses.

The two forms of hypothesis testing are based on different problem formulations. The original test is analogous to a true/false question; the Neyman-Pearson test is more like multiple choice. In the view of Tukey the former produces a conclusion on the basis of only strong evidence while the latter produces a decision on the basis of available evidence. While the two tests seem quite different both mathematically and philosophically, later developments lead to the opposite claim. Consider many tiny radioactive sources. The hypotheses become 0,1,2,3... grains of radioactive sand. There is little distinction between none or some radiation (Fisher) and 0 grains of radioactive sand versus all of the alternatives (Neyman-Pearson). The major Neyman-Pearson paper of 1933 also considered composite hypotheses (ones whose distribution includes an unknown parameter). An example proved the optimality of the (Student's) t-test, "there can be no better test for the hypothesis under consideration" (p 321). Neyman-Pearson theory was proving the optimality of Fisherian methods from its inception.

Fisher's significance testing has proven a popular flexible statistical tool in application with little mathematical growth potential. Neyman-Pearson hypothesis testing is claimed as a pillar of mathematical statistics, creating a new paradigm for the field. It also stimulated new applications in Statistical process control, detection theory, decision theory and game theory. Both formulations have been successful, but the successes have been of a different character.

The dispute over formulations is unresolved. Science primarily uses Fisher's (slightly modified) formulation as taught in introductory statistics. Statisticians study Neyman-Pearson theory in graduate school. Mathematicians are proud of uniting the formulations. Philosophers consider them separately. Learned opinions deem the formulations variously competitive (Fisher vs Neyman), incompatible or complementary. The dispute has become more complex since Bayesian inference has achieved respectability.

The terminology is inconsistent. Hypothesis testing can mean any mixture of two formulations that both changed with time. Any discussion of significance testing vs hypothesis testing is doubly vulnerable to confusion.

Fisher thought that hypothesis testing was a useful strategy for performing industrial quality control, however, he strongly disagreed that hypothesis testing could be useful for scientists. Hypothesis testing provides a means of finding test statistics used in significance testing. The concept of power is useful in explaining the consequences of adjusting the significance level and is heavily used in sample size determination. The two methods remain philosophically distinct. They usually (but *not always*) produce the same mathematical answer. The preferred answer is context dependent. While the existing merger of Fisher and Neyman-Pearson theories has been heavily criticized, modifying the merger to achieve Bayesian goals has been considered.

Criticism

Criticism of statistical hypothesis testing fills volumes citing 300–400 primary references. Much of the criticism can be summarized by the following issues:

- The interpretation of a p-value is dependent upon stopping rule and definition of multiple comparison. The former often changes during the course of a study and the latter is unavoidably ambiguous. (i.e. "p values depend on both the (data) observed and on the other possible (data) that might have been observed but weren't").
- Confusion resulting (in part) from combining the methods of Fisher and Neyman-Pearson which are conceptually distinct.
- Emphasis on statistical significance to the exclusion of estimation and confirmation by repeated experiments.
- Rigidly requiring statistical significance as a criterion for publication, resulting in publication bias. Most of the criticism is indirect. Rather than being wrong, statistical hypothesis testing is misunderstood, overused and misused.
- When used to detect whether a difference exists between groups, a paradox arises. As improvements are made to experimental design (e.g., increased precision of measurement and sample size), the test becomes more lenient. Unless one accepts the absurd assumption that all sources of noise in the data cancel out completely, the chance of finding statistical significance in either direction approaches 100%.
- Layers of philosophical concerns. The probability of statistical significance is a function of decisions made by experimenters/ analysts. If the decisions are based on convention they are termed arbitrary or mindless while those not so based may be termed subjective. To minimize type II errors, large samples are recommended. In psychology practically all null hypotheses are claimed to be false for sufficiently large samples so "...it is usually nonsensical to perform an experiment with the *sole* aim of rejecting the null hypothesis.". "Statistically significant findings are often misleading" in psychology. Statistical significance does not imply practical significance and correlation does not imply causation. Casting doubt on the null hypothesis is thus far from directly supporting the research hypothesis.
- "[I]t does not tell us what we want to know". Lists of dozens of complaints are available.

Critics and supporters are largely in factual agreement regarding the characteristics of null hypothesis significance testing (NHST): While it can provide critical information, it is *inadequate as the sole tool for statistical analysis. Successfully rejecting the null hypothesis may offer no support for the research hypothesis.* The continuing controversy concerns the selection of the best statistical practices for the near-term future given the (often poor) existing practices. Critics would prefer to ban NHST completely, forcing a complete departure from those practices, while supporters suggest a less absolute change.

Controversy over significance testing, and its effects on publication bias in particular, has produced several results. The American Psychological Association has strengthened its statistical reporting requirements after review, medical journal publishers have recognised the obligation to publish some results that are not statistically significant to combat publication bias and a journal (*Journal of Articles in Support of the Null Hypothesis*) has been created to publish such results exclusively. Textbooks have added some cautions and increased coverage of the tools necessary to estimate the size of the sample required to produce significant results. Major organisations have not abandoned use of significance tests although some have discussed doing so.

Alternatives

The numerous criticisms of significance testing do not lead to a single alternative. A unifying position of critics is that statistics should not lead to a conclusion or a decision but to a probability or to an estimated value with a confidence interval rather than to an accept-reject decision regarding a particular hypothesis. It is unlikely that the controversy surrounding significance testing will be resolved in the near future. Its supposed flaws and unpopularity do not eliminate the need for an objective and transparent means of reaching conclusions regarding studies that produce statistical results. Critics have not unified around an alternative. Other forms of reporting confidence or uncertainty could probably grow in popularity. One strong critic of significance testing suggested a list of reporting alternatives: effect sizes for importance, prediction intervals for confidence, replications and extensions for replicability, meta-analyses for generality. None of these suggested alternatives produces a conclusion/decision. Lehmann said that hypothesis testing theory can be presented in terms of conclusions/ decisions, probabilities, or confidence intervals. "The distinction between the ... approaches is largely one of reporting and interpretation."

On one "alternative" there is no disagreement: Fisher himself said, "In relation to the test of significance, we may say that a phenomenon is experimentally demonstrable when we know how to conduct an experiment which will rarely fail to give us a statistically significant result." Cohen, an influential critic of significance testing, concurred, "... don't look for a magic alternative to NHST *[null hypothesis significance testing]* ... It doesn't exist." "... given the problems of statistical induction, we must finally rely, as have the older sciences, on replication." The "alternative" to significance testing is repeated testing. The easiest way to decrease statistical uncertainty is by obtaining more data, whether by increased sample size or by repeated tests. Nickerson claimed to have never seen the publication of a literally replicated experiment in psychology. An indirect approach to replication is meta-analysis.

Bayesian inference is one proposed alternative to significance testing. (Nickerson cited 10 sources suggesting it, including Rozeboom (1960)). For example, Bayesian parameter estimation can provide rich information about the data from which researchers can draw inferences, while using uncertain priors that exert only minimal influence on the results when enough data is available. Psychologist Kruschke, John K. has suggested Bayesian estimation as an alternative for the t-test. Alternatively two competing models/hypothesis can be compared using Bayes factors. Bayesian methods could be criticized for requiring information that is seldom available in the cases where significance testing is most heavily used. Neither the prior probabilities nor the probability distribution of the test statistic under the alternative hypothesis are often available in the social sciences.

Advocates of a Bayesian approach sometimes claim that the goal of a researcher is most often to objectively assess the probability that a hypothesis is true based on the data they have collected. Neither Fisher's significance testing, nor Neyman-Pearson hypothesis testing can provide this information, and do not claim to. The probability a hypothesis is true can only be derived from use of Bayes' Theorem, which was unsatisfactory to both the Fisher and Neyman-Pearson camps due to the explicit use of subjectivity in the form of the prior probability. Fisher's strategy is to sidestep this with the p-value (an objective *index* based on the data alone) followed by *inductive inference*, while Neyman-Pearson devised their approach of *inductive behaviour*.

Philosophy

Hypothesis testing and philosophy intersect. Inferential statistics, which includes hypothesis testing, is applied probability. Both probability and its application are intertwined with philosophy. Philosopher David Hume wrote, "All knowledge degenerates into probability." Competing practical definitions of probability reflect philosophical differences. The most common application of hypothesis testing is in the scientific interpretation of experimental data, which is naturally studied by the philosophy of science.

Fisher and Neyman opposed the subjectivity of probability. Their views contributed to the objective definitions. The core of their historical disagreement was philosophical.

Many of the philosophical criticisms of hypothesis testing are discussed by statisticians in other contexts, particularly correlation does not imply causation and the design of experiments. Hypothesis testing is of continuing interest to philosophers.

Education

Statistics is increasingly being taught in schools with hypothesis testing being one of the elements taught. Many conclusions reported in the popular press (political opinion polls to medical studies) are based on statistics. An informed public should understand the limitations of statistical conclusions and many college fields of study require a course in statistics for the same reason. An introductory college statistics class places much emphasis on hypothesis testing – perhaps half of the course. Such fields as literature and divinity now include findings based on statistical analysis. An introductory statistics class teaches hypothesis testing as a cookbook process. Hypothesis testing is also taught at the postgraduate level. Statisticians learn how to create good statistical test procedures (like z, Student's t, F and chi-squared). Statistical hypothesis testing is considered a mature area within statistics, but a limited amount of development continues.

The cookbook method of teaching introductory statistics leaves no time for history, philosophy or controversy. Hypothesis testing has been taught as received unified method. Surveys showed that graduates of the class were filled with philosophical misconceptions (on all aspects of statistical inference) that persisted among instructors. While the problem was addressed more than a decade ago, and calls for educational reform continue, students still graduate from statistics classes holding fundamental misconceptions about hypothesis testing. Ideas for improving the teaching of hypothesis testing include encouraging students to search for statistical errors in published papers, teaching the history of statistics and emphasizing the controversy in a generally dry subject.

Hidden Markov Model

A hidden Markov model (HMM) is a statistical Markov model in which the system being modelled is assumed to be a Markov process with unobserved (*hidden*) states. An HMM can be presented as the simplest dynamic Bayesian network. The mathematics behind the HMM was developed by L. E. Baum and coworkers. It is closely related to an earlier work on optimal nonlinear filtering problem by Ruslan L. Stratonovich, who was the first to describe the forward-backward procedure.

In simpler Markov models (like a Markov chain), the state is directly visible to the observer, and therefore the state transition probabilities are the only parameters. In a *hidden* Markov model, the state is not directly visible, but output, dependent on the state, is visible. Each state has a probability distribution over the possible output tokens. Therefore the sequence of tokens generated by an HMM gives some information about the sequence of states. Note that the adjective 'hidden' refers to the state sequence through which the model passes, not to the parameters of the model; the model is still referred to as a 'hidden' Markov model even if these parameters are known exactly.

Hidden Markov models are especially known for their application in temporal pattern recognition such as speech, handwriting, gesture recognition, part-of-speech tagging, musical score following, partial discharges and bioinformatics.

A hidden Markov model can be considered a generalization of a mixture model where the hidden variables (or latent variables), which control the mixture component to be selected for each observation, are related through a Markov process rather than independent of each other. Recently, hidden Markov models have been generalized to pairwise Markov models and triplet Markov models which allow consideration of more complex data structures and the modelling of nonstationary data.

Description in Terms of Urns

In its discrete form, a hidden Markov process can be visualized as a generalization of the Urn problem (where each item from the urn is returned to the original urn before the next step). Consider this example: in a room that is not visible to an observer there is a genie. The room contains urns X1, X2, X3, ... each of which contains a known mix of balls, each ball labelled y1, y2, y3, ... The genie chooses an urn in that room and randomly draws a ball from that urn. It then puts the ball onto a conveyor belt, where the observer can observe the sequence of the balls but not the sequence of urns from which they were drawn. The genie has some procedure to choose urns; the choice of the urn for the n-th ball depends only upon a random number and the choice of the urn for the (n " 1)-th ball. The choice of urn does not directly depend on the urns chosen before this single previous urn; therefore, this is called a Markov process.

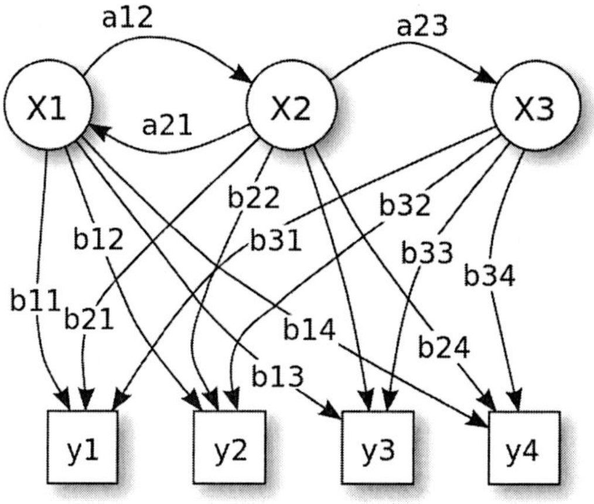

Figure: *Probabilistic parameters of a hidden Markov model (example)*

x — *states*

y — *possible observations*

a — *state transition probabilities*

b — *output probabilities*

The Markov process itself cannot be observed, and only the sequence of labelled balls can be observed, thus this arrangement is called a "hidden Markov process". This is illustrated by the lower part of the diagram shown in Figure, where one can see that balls y1, y2, y3, y4 can be drawn at each state. Even if the observer knows the composition of the urns and has just observed a sequence of three balls, *e.g.* y1, y2 and y3 on the conveyor belt, the observer still cannot be *sure* which urn (*i.e.*, at which state) the genie has drawn the third ball from. However, the observer can work out other information, such as the likelihood that the third ball came from each of the urns.

Architecture

The diagram below shows the general architecture of an instantiated HMM. Each oval shape represents a random variable that can adopt any of a number of values. The random variable $x(t)$ is the hidden state at time t (with the model from the above diagram, $x(t) \in \{x_1, x_2, x_3\}$). The random variable $y(t)$ is the observation at time t (with $y(t) \in \{y_1, y_2, y_3, y_4\}$). The arrows in the diagram (often called a trellis diagram) denote conditional dependencies.

From the diagram, it is clear that the conditional probability distribution of the hidden variable $x(t)$ at time t, given the values of the hidden variable x at all times, depends *only* on the value of the hidden variable $x(t-1)$: the values at time $t-2$ and before have no influence. This is called the Markov property. Similarly, the value of the observed variable $y(t)$ only depends on the value of the hidden variable $x(t)$ (both at time t).

In the standard type of hidden Markov model considered here, the state space of the hidden variables is discrete, while the observations themselves can either be discrete (typically generated from a categorical distribution) or continuous (typically from a Gaussian distribution). The parameters of a hidden Markov model are of two types, *transition probabilities* and *emission probabilities* (also known as *output probabilities*). The transition probabilities control the way the hidden state at time t is chosen given the hidden state at time

The hidden state space is assumed to consist of one of N possible values, modelled as a categorical distribution. This means that for each of the N possible states that a hidden variable at time can be in, there is a transition probability from this state to each of the N possible states of the hidden variable at time $t+1$, for a total of N^2 transition probabilities. Note that the set of transition probabilities for transitions from any given state must sum to 1. Thus, the $N \times N$ matrix of transition probabilities is a Markov matrix. Because any one transition probability can be determined once the others are known, there are a total of $N(N-1)$ transition parameters.

In addition, for each of the N possible states, there is a set of emission probabilities governing the distribution of the observed variable at a particular time given the state of the hidden variable at that time. The size of this set depends on the nature of the observed variable. For example, if the observed variable is discrete with M possible values, governed by a categorical distribution, there will be $M-1$ separate parameters, for a total of $N(M-1)$ emission parameters over all hidden states. On the other hand, if the observed variable is an M-dimensional vector distributed according to an arbitrary multivariate Gaussian distribution, there will be M parameters controlling the means and $M(M+1)/2$ parameters controlling the covariance matrix, for a total of

$$N(M + \frac{M(M+1)}{2}) = NM(M+3)/2 = O(NM^2)$$

emission parameters. (In such a case, unless the value of M is small, it may be more practical to restrict the nature of the covariances between individual elements of the observation vector, e.g. by assuming that the elements are independent of each other, or less restrictively, are independent of all but a fixed number of adjacent elements.)

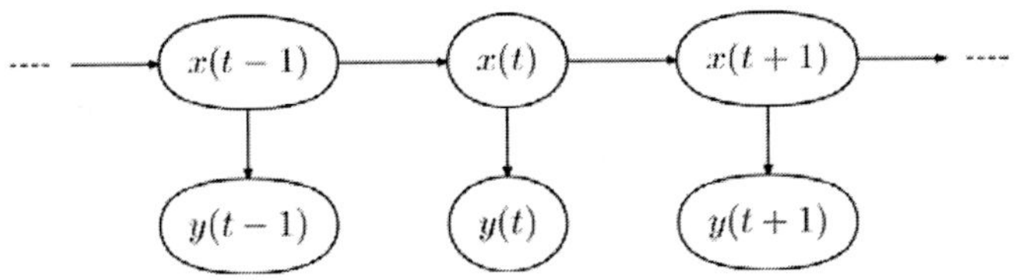

Inference

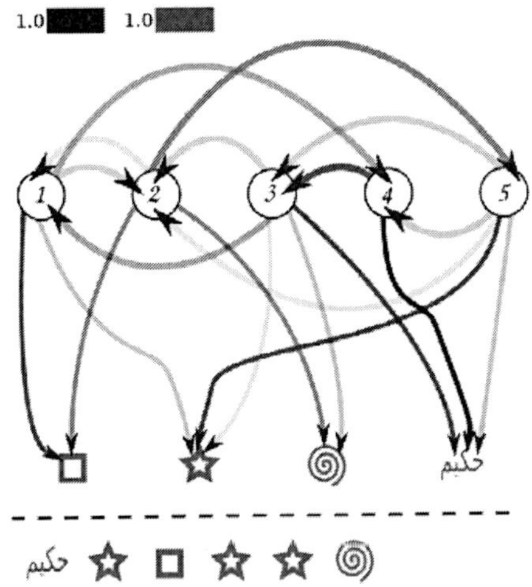

Figure: *The state transition and output probabilities of an HMM are indicated by the line opacity in the upper part of the diagram. Given that we have observed the output sequence in the lower part of the diagram, we may be interested in the most likely sequence of states that could have produced it. Based on the arrows that are present in the diagram, the following state sequences are candidates:*

5 3 2 5 3 2

4 3 2 5 3 2

3 1 2 5 3 2

We can find the most likely sequence by evaluating the joint probability of both the state sequence and the observations for each case (simply by multiplying the probability values, which here correspond to the opacities of the arrows involved). In general, this type of problem (i.e. finding the most likely explanation for an observation sequence) can be solved efficiently using the Viterbi algorithm.

Several inference problems are associated with hidden Markov models, as outlined below.

Probability of an Observed Sequence

The task is to compute, given the parameters of the model, the probability of a particular output sequence. This requires summation over all possible state sequences:

The probability of observing a sequence

$$Y = y(0), y(1), \ldots, y(L-1)$$

of length L is given by

$$P(Y) = \sum_X P(Y \mid X)P(X),$$

where the sum runs over all possible hidden-node sequences

$$X = x(0), x(1), \ldots, x(L-1).$$

Applying the principle of dynamic programming, this problem, too, can be handled efficiently using the forward algorithm.

Probability of the Latent Variables

A number of related tasks ask about the probability of one or more of the latent variables, given the model's parameters and a sequence of observations $y(1), \ldots, y(t)$.

Filtering

The task is to compute, given the model's parameters and a sequence of observations, the distribution over hidden states of the last latent variable at the end of the sequence, i.e. to compute $P(x(t) \mid y(1), \ldots, y(t))$. This task is normally used when the sequence of latent variables is thought of as the underlying states that a process moves through at a sequence of points of time, with corresponding observations at each point in time. Then, it is natural to ask about the state of the process at the end.

This problem can be handled efficiently using the forward algorithm.

Smoothing

This is similar to filtering but asks about the distribution of a latent variable somewhere in the middle of a sequence, i.e. to compute $P(x(k) \mid y(1), \ldots, y(t))$ for some $k < t$. From the perspective described above, this can be thought of as the probability distribution over hidden states for a point in time k in the past, relative to time t.

The forward-backward algorithm is an efficient method for computing the smoothed values for all hidden state variables.

Most Likely Explanation

The task, unlike the previous two, asks about the joint probability of the *entire* sequence of hidden states that generated a particular sequence of observations. This task is generally applicable when HMM's are applied to different sorts of problems from those for which the tasks of filtering and smoothing are applicable. An example is part-of-speech tagging, where the hidden states represent the underlying parts of speech corresponding to an observed sequence of words. In this case, what is of interest is the entire sequence of parts of speech, rather than simply the part of speech for a single word, as filtering or smoothing would compute.

This task requires finding a maximum over all possible state sequences, and can be solved efficiently by the Viterbi algorithm.

Statistical Significance

For some of the above problems, it may also be interesting to ask about statistical significance. What is the probability that a sequence drawn from some null distribution will have an HMM probability (in the case of the forward algorithm) or a maximum state sequence probability (in the case of the Viterbi algorithm) at least as large as that of a particular output sequence? When an HMM is used to evaluate the relevance of a hypothesis for a particular output sequence, the statistical significance indicates the false positive rate associated with failing to reject the hypothesis for the output sequence.

A Concrete Example

Consider two friends, Alice and Bob, who live far apart from each other and who talk together daily over the telephone about what they did that day. Bob is only interested in three activities: walking in the park, shopping, and cleaning his apartment. The choice of what to do is determined exclusively by the weather on a given day. Alice has no definite information about the weather where Bob lives, but she knows general trends. Based on what Bob tells her he did each day, Alice tries to guess what the weather must have been like.

Alice believes that the weather operates as a discrete Markov chain. There are two states, "Rainy" and "Sunny", but she cannot observe them directly, that is, they are *hidden* from her. On each day, there is a certain chance that Bob will perform one of the following activities, depending on the weather: "walk", "shop", or "clean". Since Bob tells Alice about his activities, those are the *observations*. The entire system is that of a hidden Markov model (HMM).

Alice knows the general weather trends in the area, and what Bob likes to do on average. In other words, the parameters of the HMM are known. They can be represented as follows in the Python programming language:

```
states = ('Rainy', 'Sunny')
observations = ('walk', 'shop', 'clean')
start_probability = {'Rainy': 0.6, 'Sunny': 0.4}
transition_probability = {
   'Rainy' : {'Rainy': 0.7, 'Sunny': 0.3},
   'Sunny' : {'Rainy': 0.4, 'Sunny': 0.6},
   }

emission_probability = {
   'Rainy' : {'walk': 0.1, 'shop': 0.4, 'clean': 0.5},
   'Sunny' : {'walk': 0.6, 'shop': 0.3, 'clean': 0.1},
 }
```

In this piece of code, `start_probability` represents Alice's belief about which state the HMM is in when Bob first calls her (all she knows is that it tends to be rainy on average).

The particular probability distribution used here is not the equilibrium one, which is (given the transition probabilities) approximately `{'Rainy': 0.57, 'Sunny': 0.43}`. The `transition_probability` represents the change of the weather in the underlying Markov chain. In this example, there is only a 30% chance that tomorrow will be sunny if today is rainy.

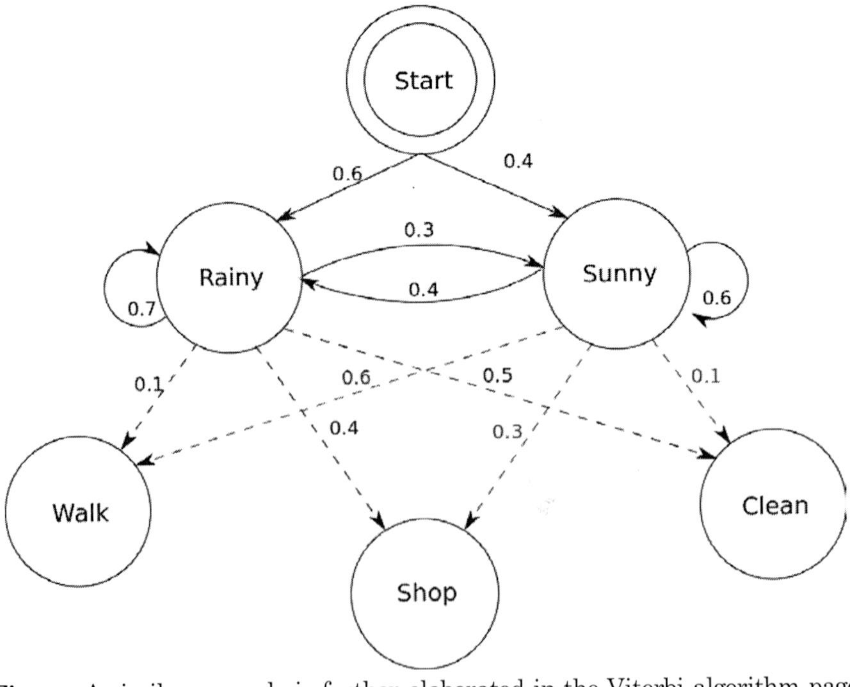

Figure: A similar example is further elaborated in the Viterbi algorithm page.

The `emission_probability` represents how likely Bob is to perform a certain activity on each day. If it is rainy, there is a 50% chance that he is cleaning his apartment; if it is sunny, there is a 60% chance that he is outside for a walk.

Learning

The parameter learning task in HMMs is to find, given an output sequence or a set of such sequences, the best set of state transition and output probabilities. The task is usually to derive the maximum likelihood estimate of the parameters of the HMM given the set of output sequences. No tractable algorithm is known for solving this problem exactly, but a local maximum likelihood can be derived efficiently using the Baum–Welch algorithm or the Baldi–Chauvin algorithm.

The Baum–Welch algorithm is a special case of the expectation-maximization algorithm.

Mathematical Description

General Description

A basic, non-Bayesian hidden Markov model can be described as follows:

N	=	number of states
T	=	number of observations
$\theta_{i=1\ldots N}$	=	emission parameter for an observation associated with state i
$\phi_{i=1\ldots N, j=1\ldots N}$	=	probability of transition from state i to state j
$\phi_{i=1\ldots N}$	=	N -dimensional vector, composed of $\phi_{i,1\ldots N}$; must sum to 1
$x_{t=1\ldots T}$	=	state of observation at time t
$y_{t=1\ldots T}$	=	observation at time t
$F(y \mid \theta)$	=	probability distribution of an observation, parametrized on θ
$x_{t=2\ldots T}$	=	$\text{Categorical}(\phi_{x_{t-1}})$
$y_{t=1\ldots T}$	=	$F(\theta_{x_t})$

Note that, in the above model (and also the one below), the prior distribution of the initial state x_1 is not specified. Typical learning models correspond to assuming a discrete uniform distribution over possible states (i.e. no particular prior distribution is assumed).

In a Bayesian setting, all parameters are associated with random variables, as follows:

N,T	=	as above
$\theta_{i=1...N}, \phi_{i=1...N,j=1...N}, \phi_{i=1...N}$	=	as above
$x_{t=1...T}, y_{t=1...T}, F(y\mid\theta)$	=	as above
α	=	shared hyperparameter for emission parameters
β	=	shared hyperparameter for transition parameters
$H(\theta\mid\alpha)$	=	prior probability distribution of emission parameters, parametrized on α
$\theta_{i=1...N}$	=	$H(\alpha)$
$\phi_{i=1...N}$	=	Symmetric- $\text{Dirichlet}_N(\beta)$
$x_{t=2...T}$	=	$\text{Categorical}(\phi_{x_{t-1}})$
$y_{t=1...T}$	=	$F(\theta_{x_t})$

These characterizations use F and H to describe arbitrary distributions over observations and parameters, respectively.

Typically H will be the conjugate prior of F. The two most common choices of F are Gaussian and categorical.

Compared with a Simple Mixture Model

As mentioned above, the distribution of each observation in a hidden Markov model is a mixture density, with the states of the corresponding to mixture components.

It is useful to compare the above characterizations for an HMM with the corresponding characterizations, of a mixture model, using the same notation.

A non-Bayesian mixture model:

N	=	number of mixture components
T	=	number of observations
$\theta_{i=1...N}$	=	parameter of distribution of observation associated with component i
$\phi_{i=1...N}$	=	mixture weight, i.e. prior probability of component i
ϕ	=	N-dimensional vector, composed of $\phi_{1...N}$; must sum to 1
$x_{t=1...T}$	=	component of observation t
$y_{t=1...T}$	=	observation t
$F(y\mid\theta)$	=	probability distribution of an observation, parametrized on θ
$x_{t=1...T}$	=	$\text{Categorical}(\phi)$
$y_{t=1...T}$	=	$F(\theta_{x_t})$

A Bayesian mixture model:

N,T	=	as above
$\theta_{i=1...N}, \phi_{i=1...N}, \phi$	=	as above
$x_{t=1...T}, y_{t=1...T}, F(y \mid \theta)$	=	as above
α	=	shared hyperparameter for component parameters
β	=	shared hyperparameter for mixture weights
$H(\theta \mid \alpha)$	=	prior probability distribution of component parameters, parametrized on α
$\theta_{i=1...N}$	=	$H(\alpha)$
ϕ	=	$\text{Symmetric-Dirichlet}_N(\beta)$
$x_{t=1...T}$	=	$\text{Categorical}(\phi)$
$y_{t=1...T}$	=	$F(\theta_{x_t})$

A Two-Level Bayesian HMM

An alternative for the above two Bayesian examples would be to add another level of prior parameters for the transition matrix. That is, replace the lines with the following:

β	=	concentration hyperparameter controlling the density of the transition matrix
$\phi_{i=1..N}$	~	$\text{Symmetric-Dirichlet}_N(\beta)$

γ	=	concentration hyperparameter controlling how many states are intrinsically likely
β	=	concentration hyperparameter controlling the density of the transition matrix
η	=	N-dimensional vector of probabilities, specifying the intrinsic probability of a given state
η	=	$\text{Symmetric-Dirichlet}_N(\gamma)$
$\phi_{i=1..N}$	=	$\text{Dirichlet}_N(\beta N \eta)$

What this means is the following:

1. n is a probability distribution over states, specifying which states are inherently likely. The greater the probability of a given state in this vector, the more likely is a transition to that state (regardless of the starting state).
2. γ controls the density of η. Values significantly above 1 cause a dense vector where all states will have similar prior probabilities. Values significantly below 1 cause a sparse vector where only a few states are inherently likely (have prior probabilities significantly above 0).
3. β controls the density of the transition matrix, or more specifically, the density of the N different probability vectors $\phi_{i=1...N}$ specifying the probability of transitions out of state i to any other state.

Imagine that the value of β is significantly above 1. Then the different ϕ vectors will be dense, i.e. the probability mass will be spread out fairly evenly over all states. However, to the extent that this mass is unevenly spread, $\boldsymbol{\eta}$ controls which states are likely to get more mass than others.

Now, imagine instead that β is significantly below 1. This will make the ϕ vectors sparse, i.e. almost all the probability mass is distributed over a small number of states, and for the rest, a transition to that state will be very unlikely. Notice that there are different ϕ vectors for each starting state, and so even if all the vectors are sparse, different vectors may distribute the mass to different ending states. However, for all of the vectors, controls which ending states are likely to get mass assigned to them.

For example, $\boldsymbol{\eta}$ if is 0.1, then each ϕ will be sparse and, for any given starting state i, the set of states $\mathbf{J}_i$ to which transitions are likely to occur will be very small, typically having only one or two members. Now, if the probabilities in $\boldsymbol{\eta}$ are all the same (or equivalently, one of the above models without $\boldsymbol{\eta}$ is used), then for different i, there will be different states in the corresponding $\mathbf{J}_i$, so that all states are equally likely to occur in any given $\mathbf{J}_i$. On the other hand, if the values in $\boldsymbol{\eta}$ are unbalanced, so that one state has a much higher probability than others, almost all $\mathbf{J}_i$ will contain this state; hence, regardless of the starting state, transitions will nearly always occur to this given state.

Hence, a two-level model such as just described allows independent control over (1) the overall density of the transition matrix, and (2) the density of states to which transitions are likely (i.e. the density of the prior distribution of states in any particular hidden variable x_i).

In both cases this is done while still assuming ignorance over which particular states are more likely than others. If it is desired to inject this information into the model, the probability vector $\boldsymbol{\eta}$ can be directly specified; or, if there is less certainty about these relative probabilities, a non-symmetric Dirichlet distribution can be used as the prior distribution over $\boldsymbol{\eta}$.

That is, instead of using a symmetric Dirichlet distribution with a single parameter γ (or equivalently, a general Dirichlet with a vector all of whose values are equal to γ), use a general Dirichlet with values that are variously greater or less than γ, according to which state is more or less preferred.

Applications

HMMs can be applied in many fields where the goal is to recover a data sequence that is not immediately observable (but other data that depend on the sequence are). Applications include:

- Single Molecule Kinetic analysis
- Cryptanalysis
- Speech recognition
- Speech synthesis
- Part-of-speech tagging
- Document Separation in scanning solutions
- Machine translation
- Partial discharge
- Gene prediction
- Alignment of bio-sequences
- Time Series Analysis
- Human Activity recognition
- Protein folding
- Metamorphic Virus Detection
- DNA Motif Discovery

History

The forward and backward recursions used in HMM as well as computations of marg eld of bioinformatics.

Types

Hidden Markov models can model complex Markov processes where the states emit the observations according to some probability distribution. One such example of distribution is Gaussian distribution, in such a Hidden Markov Model the states output is represented by a Gaussian distribution.

Moreover it could represent even more complex behaviour when the output of the states is represented as mixture of two or more Gaussians, in which case the probability of generating an observation is the product of the probability of first selecting one of the Gaussians and the probability of generating that observation from that Gaussian.

Extensions

In the hidden Markov models considered above, the state space of the hidden variables is discrete, while the observations themselves can either be discrete (typically generated from a categorical distribution) or continuous (typically from a Gaussian distribution). Hidden Markov models can also be generalized to allow continuous state spaces. Examples of such models are those where the Markov process over hidden variables is a linear dynamical system, with a linear relationship among related variables and where all hidden and observed variables follow a Gaussian distribution. In simple cases, such as the linear dynamical system just mentioned, exact inference is tractable (in this case, using the Kalman filter); however, in general, exact inference in HMMs with continuous latent variables is infeasible, and approximate methods must be used, such as the extended Kalman filter or the particle filter.

Hidden Markov models are generative models, in which the joint distribution of observations and hidden states, or equivalently both the prior distribution of hidden states (the *transition probabilities*) and conditional distribution of observations given states (the *emission probabilities*), is modelled. The above algorithms implicitly assume a uniform prior distribution over the transition probabilities. However, it is also possible to create hidden Markov models with other types of prior distributions. An obvious candidate, given the categorical distribution of the transition probabilities, is the Dirichlet distribution, which is the conjugate prior distribution of the categorical distribution. Typically, a symmetric Dirichlet distribution is chosen, reflecting ignorance about which states are inherently more likely than others. The single parameter of this distribution (termed the *concentration parameter*) controls the relative density or sparseness of the resulting transition matrix.

A choice of 1 yields a uniform distribution. Values greater than 1 produce a dense matrix, in which the transition probabilities between pairs of states are likely to be nearly equal. Values less than 1 result in a sparse matrix in which, for each given source state, only a small number of destination states have non-negligible transition probabilities. It is also possible to use a two-level prior Dirichlet distribution, in which one Dirichlet distribution (the upper distribution) governs the parameters of another Dirichlet distribution (the lower distribution), which in turn governs the transition probabilities. The upper distribution governs the overall distribution of states, determining how likely each state is to occur; its concentration parameter determines the density or sparseness of states. Such a two-level prior distribution, where both concentration parameters are set to produce sparse distributions, might be useful for example in unsupervised part-of-speech tagging, where some parts of speech occur much more commonly than others; learning algorithms that assume a uniform prior distribution generally perform poorly on this task. The parameters of models of this sort, with non-uniform prior distributions, can be learned using Gibbs sampling or extended versions of the expectation-maximization algorithm.

An extension of the previously described hidden Markov models with Dirichlet priors uses a Dirichlet process in place of a Dirichlet distribution. This type of model allows for an unknown and potentially infinite number of states. It is common to use a two-level Dirichlet process, similar to the previously described model with two levels of Dirichlet distributions. Such a model is called a *hierarchical Dirichlet process hidden Markov model*, or *HDP-HMM* for short. It was originally described under the name "Infinite Hidden Markov Model" and was further formalized in.

A different type of extension uses a discriminative model in place of the generative model of standard HMMs. This type of model directly models the conditional distribution of the hidden states given the observations, rather than modelling the joint distribution. An example of this model is the so-called *maximum entropy Markov model* (MEMM), which models the conditional distribution of the states using logistic regression (also known as a "maximum entropy model"). The advantage of this type of model is that arbitrary features (i.e. functions) of the observations can be modelled, allowing domain-specific knowledge of the problem at hand to be injected into the model.

Models of this sort are not limited to modelling direct dependencies between a hidden state and its associated observation; rather, features of nearby observations, of combinations of the associated observation and nearby observations, or in fact of arbitrary observations at any distance from a given hidden state can be included in the process used to determine the value of a hidden state. Furthermore, there is no need for these features to be statistically independent of each other, as would be the case if such features were used in a generative model. Finally, arbitrary features over pairs of adjacent hidden states can be used rather than simple transition probabilities. The disadvantages of such models are:

(1) The types of prior distributions that can be placed on hidden states are severely limited;

(2) It is not possible to predict the probability of seeing an arbitrary observation.

This second limitation is often not an issue in practice, since many common usages of HMM's do not require such predictive probabilities.

A variant of the previously described discriminative model is the linear-chain conditional random field. This uses an undirected graphical model (aka Markov random field) rather than the directed graphical models of MEMM's and similar models. The advantage of this type of model is that it does not suffer from the so-called *label bias* problem of MEMM's, and thus may make more accurate predictions. The disadvantage is that training can be slower than for MEMM's.

Yet another variant is the *factorial hidden Markov model*, which allows for a single observation to be conditioned on the corresponding hidden variables of a set of K independent Markov chains, rather than a single Markov chain. It is equivalent to a single HMM, with N^K states (assuming there are N states for each chain), and therefore, learning in such a model is difficult: for a sequence of length T, a straightforward Viterbi algorithm has complexity $O(N^{2K}T)$. To find an exact solution, a junction tree algorithm could be used, but it results in an $O(N^{K+1}KT)$ complexity. In practice, approximate techniques, such as variational approaches, could be used.

All of the above models can be extended to allow for more distant dependencies among hidden states, e.g. allowing for a given state to be dependent on the previous two or three states rather than a single previous state; i.e. the transition probabilities are extended to encompass sets of three or four adjacent states (or in general K adjacent states). The disadvantage of such models is that dynamic-programming algorithms for training them have an $O(N^K T)$ running time, for K adjacent states and T total observations (i.e. a length-T Markov chain).

Another recent extension is the *triplet Markov model*, in which an auxiliary underlying process is added to model some data specificities. Many variants of this model have been proposed. One should also mention the interesting link that has been established between the *theory of evidence* and the *triplet Markov models* and which allows to fuse data in Markovian context and to model nonstationary data.

Information Processing Theory

The information processing theory approach to the study of cognitive development evolved out of the American experimental tradition in psychology. Developmental psychologists who adopt the information-processing perspective account for mental development in terms of maturational changes in basic components of a child's mind. The theory is based on the idea that humans process the information they receive, rather than merely responding to stimuli. This perspective equates the mind to a computer, which is responsible for analyzing information from the environment. According to the standard information-processing model for mental development, the mind's machinery includes attention mechanisms for bringing information in, working memory for actively manipulating information, and long-term memory for passively holding information so that it can be used in the future. This theory addresses how as children grow, their brains likewise mature, leading to advances in their ability to process and respond to the information they received through their senses. The theory emphasizes a continuous pattern of development, in contrast with Cognitive Developmental theorists such as Jean Piaget that thought development occurred in stages at a time.

Emergence

Beginning in the 1950s, a major change occurred in the field of Psychology that has come to be known as the Cognitive Revolution. The cognitive revolution took form as what is now known as Cognitive Psychology. This field of psychology had freed itself from the behaviourist views that were dominant in the 1950s. It wanted to look at the interior mental processes, rather than the observable exterior views that behaviourism held. This revolution had a huge impact on theory and research in the field of psychology, as well as many other disciplines, such as human-computer interaction, human factors and ergonomics. Overall, information-processing models helped reestablish internal thought processes as a legitimate area of scientific research.

A central metaphor that was adopted by cognitivists at this time was the computer, which served to provide these researchers important clues and directions in understanding the human brain and how it processes information. Many psychologists and researchers believe that the Information Processing Theory was influenced by computers, in that the human mind is similar to a computer. However, today the metaphor of mind as computer has faded. The analogy has many strengths, in that humans have different memory stores and information is transferred from one store to another, however it does little to actually explain how the process works and has thus diminished in popularity.

Human as Computer

Within this model, humans are routinely compared to computers. This comparison is used as a means of better understanding the way information is processed and stored in the human mind. Therefore, when analyzing what actually develops within this model, the more specific comparison is between the human brain and computers. Computers were introduced to the study of development and provided a new way of studying intelligence (Lachman, 1979) and added further legitimacy to the scientific study of the mind (Goodwin, 2005, p. 411). In the model below, you can see the direct comparison between human processing and computer processing. Within this model, information is taken in (or input). Information is encoded to give meaning and compared with stored information. If a person is working on a task, this is where the working memory is enacted. An example of that for a computer is the CPU. In both cases, information is encoded, given meaning, and combined with previously stored information to enact the task. The latter step is where the information is stored where it can later be retrieved when needed. For computers, this would be akin to saving information on a hard drive, where you would then upload the saved data when working on a future task (using your working memory as in step 2).

Cognitive Processes

Cognitive processes include perception, recognition, imagining, remembering, thinking, judging, reasoning, problem solving, conceptualizing and singing, planning and more terms and applications. These cognitive processes can emerge from human language, thought, imagery and symbols.

Out of all of these specific cognitive processes, many cognitive psychologists study language-acquisition, altered states of mind and consciousness, visual perception, auditory perception, short-term memory, long-term memory, storage, retrieval, perceptions of thought and much more.

The Four Pillars of the Information Processing Model

There are four fundamental assumptions – or four pillars – of the information processing approach. These pillars underlay and support this approach, as well as many other cognitive models.

- Thinking: The process of thinking includes the activities of perception of external stimuli, encoding the same and storing the data so perceived and encoded in one's mental recesses.
- Analysis of stimuli: This is the process by which the encoded stimuli are altered to suit the brain's cognition and interpretation process to enable decision making. There are four distinct sub-processes that form a favourable alliance to make

the brain arrive at a conclusion regarding the encoded stimuli it has received and kept stored. These four sub-processes are encoding, strategization, generalization and automatization.

- Situational modification: This is the process by which an individual uses his experience, which is nothing other than a collection of stored memories, to handle a similar situation in future. In case of certain differences in both situations, the individual modifies the decisions they took during their previous experience to come up with solutions for the somewhat different problem.
- Obstacle evaluation: This step maintains that besides the subject's individual development level, the nature of the obstacle or problem should also be taken into consideration while evaluating the subject's intellectual, problem solving and cognitive acumen. Sometimes, unnecessary and misleading information can confuse the subject who may show signs of confusion while dealing with a situation similar to the one exposed to before and handled successfully.

Structure of the Information-Processing System

The standard information-processing model has three major components: sensory register, short-term memory (working memory), and long-term memory.

Sensory Register

Âàch sensory system hàs its own sensory store, or sensory register, which receives and briefly holds all of the external ànd internal stimuli that it receives. The sensîry register stores sensory information long enough for unconscious πrîcesses to analyze it tî determine whether the inðut should be brought into working memory, ît discàrded.

Working Memory

Working memory is believed to be the centre of conscious thought, analogous to the central processing unit of a computer, where information from long-term memory and the environment is combined to help solve problems. However, the working memory has a small capacity so that it is not able to attend to much information at a time, thereby limiting the abilities of humans to solve problems. The information processing perspective proposes that as children grow until about 15 years old, their working memory capacity for verbal/visual information also steadily increases, as demonstrated by improved performance on fluid intelligence tests. Many proponents of the information processing system correlate this increased working-memory capacity with increased speed of processing, the speed at which a person can fluently carry out relatively elementary information-processing tasks. It is believed that the physical maturation of the brain that occurs throughout childhood may cause faster processing speeds. This faster processing speed permits faster mental movement from one item of information to another, which improves one's ability to keep track of a number of different items in working memory at once.

Long-Term Memory

Long-term memory is the stored representation of all that a person knows. The items stored in long-term memory lie dormant until they are called back into the working memory and thus put to use.

Long-term memory consists of explicit and implicit long-term memory systems. Children exhibit implicit long-term memory – memories that affect behaviour, but with which we are unable to report, such as procedural memories – beginning in early infancy on. There also exist two categories of explicit memory: semantic and episodic memory. Children exhibit the ability to form semantic memories as quickly as when they learn words, which possibly aids the development of vocabulary. In contrast, episodic memories develop relatively slowly, appearing at about 3 years of age when children are able to answer questions reliably about past experiences.

Many psychologists believe that the ability to form episodic memories increases gradually throughout childhood due to continued maturation of the brain, particularly in the prefrontal lobes. Proponents of the information processing theory make sense of the development of memory systems, from implicit → semantic → episodic, in terms of childhood developmental needs.

Intelligence as Processing

Before information processing theory, psychologists had trouble operationally defining intelligence. One psychologist even ventured to define intelligence as, "What intelligence tests measure." Information processing defines intelligence as processing. In this way, intelligence can be researched with a more definitive purpose. If the better processor is the smarter human, then getting at what makes people better processors from a psychological level all the way to a biochemical level will help humans understand intelligence much better.

Nature versus Nurture

This theory views humans as being like machines, actively inputting, retrieving, processing and storing information. Context, social content, and social influences on processing are generally ignored in favour of a focus on internal systemic processes (Miller, 2011). Nature provides the hardware, or the neurological processing system likely predisposed to economical and efficient processing, as well as being pre-tuned to attend to specific stimuli. The "Nurture" component presents as the environment which provides the stimuli to be inputted and processed by the system

Quantitative versus Qualitative

Information processing theory combines elements of both quantitative and qualitative development. Qualitative development occurs through the emergence of new strategies for information storage and retrieval, developing representational abilities (such as the utilisation of language to represent concepts), or obtaining problem-solving rules (Miller, 2011). Increases in the knowledge base or the ability to remember more items in working memory are examples of quantitative changes, as well as increases in the strength of connected cognitive associations (Miller, 2011). The qualitative and quantitative components often interact together to develop new and more efficient strategies within the processing system.

Current Areas of Research

Information Processing Theory is currently being utilised in the study of computer or intelligence. This theory has also been applied to systems beyond the individual, including families and business organisations. For example, Ariel (1987) applied Information Processing Theory to family systems, with sensing, attending, and encoding stimuli occurring either within individuals within the system or as the family system itself. Unlike traditional systems theory, where the family system tends to maintain stasis and resists incoming stimuli which would violate the system's rules, the Information Processing family develops individual and mutual schemas which influence what and is attended to and processed. Dysfunctions can occur both on the individual level as well as within the family system itself, creating more targets for therapeutic change.) utilised Information Processing Theory to describe business organisational behaviour, as well as to present a model describing how effective and ineffective business strategies are developed. In their study, components of organisations that "sense" market information are identified as well as how organisations attend to this information; which gatekeepers determine what information is relevant/important for the organisation, how this is organised into the existing culture (organisational schemas), and whether or not the organisation has effective or ineffective processes for their long-term strategy.

Transcriptome

The transcriptome is the set of all RNA molecules, including mRNA, rRNA, tRNA, and other non-coding RNA transcribed in one cell or a population of cells. It differs from the exome in that it includes only those RNA molecules found in a specified cell population, and usually includes the amount or concentration of each RNA molecule in addition to the molecular identities.

Scope

The term can be applied to the total set of transcripts in a given organism, or to the specific subset of transcripts present in a particular cell type. Unlike the genome, which is roughly fixed for a given cell line (excluding mutations), the transcriptome can vary with external environmental conditions. Because it includes all *mRNA* transcripts in the cell, the transcriptome reflects the genes that are being actively expressed at any given time, with the exception of mRNA degradation phenomena such as transcriptional attenuation. The study of *transcriptomics*, also referred to as expression profiling, examines the expression level of mRNAs in a given cell population, often using high-throughput techniques based on DNA microarray technology. The use of next-generation sequencing technology to study the transcriptome at the nucleotide level is known as RNA-Seq.

Methods of Construction

There are two general methods of inferring transcriptomes. One approach maps sequence reads onto a reference genome, either of the organism itself (whose transcriptome is being studied) or of a closely related species. The other approach, *de novo* transcriptome assembly, uses software to infer transcripts directly from short sequence reads.

Analysis

A number of organism-specific transcriptome databases have been constructed and annotated to aid in the identification of genes that are differentially expressed in distinct cell populations.

RNA-seq is emerging (2013) as the method of choice for measuring transcriptomes of organisms, though the older technique of DNA microarrays is still used.

Applications

The transcriptomes of stem cells and cancer cells are of particular interest to researchers who seek to understand the processes of cellular differentiation and carcinogenesis.

Analysis of the transcriptomes of human oocytes and embryos is used to understand the molecular mechanisms and signaling pathways controlling early embryonic development, and could theoretically be a powerful tool in making proper embryo selection in vitro fertilization.

Transcriptomics is an emerging and continually growing field in biomarker discovery for use in assessing the safety of drugs or chemical risk assessment.

Relation to Proteome

The transcriptome can be seen as a precursor for the proteome, that is, the entire set of proteins expressed by a genome.

However, the analysis of relative mRNA expression levels can be complicated by the fact that relatively small changes in mRNA expression can produce large changes in the total amount of the corresponding protein present in the cell. One analysis method, known as Gene Set Enrichment Analysis, identifies coregulated gene networks rather than individual genes that are up- or down-regulated in different cell populations.

Although microarray studies can reveal the relative amounts of different mRNAs in the cell, levels of mRNA are not directly proportional to the expression level of the proteins they code for. The number of protein molecules synthesized using a given mRNA molecule as a template is highly dependent on translation-initiation features of the mRNA sequence; in particular, the ability of the translation initiation sequence is a key determinant in the recruiting of ribosomes for protein translation. The complete protein complement of a cell or organism is known as the proteome.

Chapter 7

Multivariate Analysis

Multivariate analysis (MVA) is based on the statistical principle of multivariate statistics, which involves observation and analysis of more than one statistical outcome variable at a time. In design and analysis, the technique is used to perform trade studies across multiple dimensions while taking into account the effects of all variables on the responses of interest.

Uses for multivariate analysis include:

- Design for capability (also known as capability-based design)
- Inverse design, where any variable can be treated as an independent variable
- Analysis of Alternatives (AoA), the selection of concepts to fulfill a customer need
- Analysis of concepts with respect to changing scenarios
- Identification of critical design drivers and correlations across hierarchical levels.

Multivariate analysis can be complicated by the desire to include physics-based analysis to calculate the effects of variables for a hierarchical "system-of-systems." Often, studies that wish to use multivariate analysis are stalled by the dimensionality of the problem. These concerns are often eased through the use of surrogate models, highly accurate approximations of the physics-based code. Since surrogate models take the form of an equation, they can be evaluated very quickly. This becomes an enabler for large-scale MVA studies: while a Monte Carlo simulation across the design space is difficult with physics-based codes, it becomes trivial when evaluating surrogate models, which often take the form of response surface equations.

History

Anderson's 1958 textbook, *An Introduction to Multivariate Analysis*, educated a generation of theorists and applied statisticians; Anderson's book emphasizes hypothesis testing via likelihood ratio tests and the properties of power functions: Admissibility, unbiasedness and monotonicity.

Factor Analysis

Factor analysis is a statistical method used to describe variability among observed, correlated variables in terms of a potentially lower number of unobserved variables called factors. For example, it is possible that variations in four observed variables mainly reflect the variations in two unobserved variables. Factor analysis searches for such joint variations in response to unobserved latent variables. The observed variables are modelled as linear combinations of the potential factors, plus "error" terms. The information gained about the interdependencies between observed variables can be used later to reduce the set of variables in a dataset. Computationally this technique is equivalent to low rank approximation of the matrix of observed variables. Factor analysis originated in psychometrics, and is used in behavioural sciences, social sciences, marketing, product management, operations research, and other applied sciences that deal with large quantities of data.

Factor analysis is related to principal component analysis (PCA), but the two are not identical. Latent variable models, including factor analysis, use regression modelling techniques to test hypotheses producing error terms, while PCA is a descriptive statistical technique. There has been significant controversy in the field over the equivalence or otherwise of the two techniques.

Statistical Model

Definition: Suppose we have a set of p observable random variables, $x_1,\ldots,x_p$ with means $\mu_1,\ldots,\mu_p$.

Suppose for some unknown constants l_{ij} and k unobserved random variables F_j, where $i\in 1,\ldots,p$ and $j\in 1,\ldots,k$, where $k < p$, we have

$$x_i - \mu_i = l_{i1}F_1 + \cdots + l_{ik}F_k + \varepsilon_i.$$

Here, the ε_i are independently distributed error terms with zero mean and finite variance, which may not be the same for all i. Let $\mathrm{Var}(\varepsilon_i) = \psi_i$, so that we have

$$\text{Cov}(\varepsilon) = \text{Diag}(\psi_1, \ldots, \psi_p) = \Psi \text{ and } \text{E}(\varepsilon) = 0.$$

In matrix terms, we have

$$x - \mu = LF + \varepsilon.$$

If we have n observations, then we will have the dimensions $x_{p \times n}$, $L_{p \times k}$, and $F_{k \times n}$. Each column of x and F denote values for one particular observation, and matrix L does not vary across observations.

Also we will impose the following assumptions on .

1. F and are independent.
2. $\text{E}(F) = 0$
3. $\text{Cov}(F) = I$ (to make sure that the factors are uncorrelated)

Any solution of the above set of equations following the constraints for F is defined as the *factors*, and L as the *loading matrix*.

Suppose $\text{Cov}(x - \mu) = \Sigma$. Then note that from the conditions just imposed on F, we have

$$\text{Cov}(x - \mu) = \text{Cov}(LF + \varepsilon),$$

or

$$\Sigma = L\text{Cov}(F)L^T + \text{Cov}(\varepsilon),$$

or

$$\Sigma = LL^T + \Psi.$$

Note that for any orthogonal matrix Q if we set $L = LQ$ and $F = Q^T F$, the criteria for being factors and factor loadings still hold. Hence a set of factors and factor loadings is identical only up to orthogonal transformation.

Example

The following example is for expository purposes, and should not be taken as being realistic. Suppose a psychologist proposes a theory that there are two kinds of intelligence, "verbal intelligence" and "mathematical intelligence", neither of which is directly observed. Evidence for the theory is sought in the examination scores from each of 10 different academic fields of 1000 students. If each student is chosen randomly from a large population, then each student's 10 scores are random variables. The psychologist's theory may say that for each of the 10 academic fields, the score averaged over the group of all students who share some common pair of values for verbal and mathematical "intelligences" is some constant times their level of verbal intelligence plus another constant times their level of mathematical intelligence, i.e., it is a combination of those two "factors". The numbers for a particular subject, by which the two kinds of intelligence are multiplied to obtain the expected score, are posited by the theory to be the same for all intelligence level pairs, and are called "factor loadings" for this subject. For example, the theory may hold that the average student's aptitude in the field of taxonomy is:

{10 × the student's verbal intelligence} + {6 × the student's mathematical intelligence}.

The numbers 10 and 6 are the factor loadings associated with taxonomy. Other academic subjects may have different factor loadings.

Two students having identical degrees of verbal intelligence and identical degrees of mathematical intelligence may have different aptitudes in taxonomy because individual aptitudes differ from average aptitudes. That difference is called the "error" — a statistical term that means the amount by which an individual differs from what is average for his or her levels of intelligence.

The observable data that go into factor analysis would be 10 scores of each of the 1000 students, a total of 10,000 numbers. The factor loadings and levels of the two kinds of intelligence of each student must be inferred from the data.

Mathematical Model of the Same Example

In the following, matrices will be indicated by indexed variables. "Subject" indices will be indicated using letters a,b and c, with values running from 1 to N_a which is equal to 10 in the above example. "Factor" indices will be indicated using letters p, q and r, with values running from 1 to N_p which is equal to 2 in the above example. "Instance" or "sample" indices will be indicated using letters i,j and k, with values running from 1 to N_i. In the example above, if a sample of $N_i = 1000$ students responded to the $N_a = 10$ questions, the ith student's score for the ath question are given by x_{ai}. The purpose of factor analysis is to characterize the correlations between the variables x_a of which the x_{ai} are a particular instance, or set of observations. In order that the variables be on equal footing, they are standardized:

$$z_{ai} = \frac{x_{ai} - \mu_a}{\sigma_a}$$

where the sample mean is:

$$\mu_a = \frac{1}{N}\sum_i x_{ai}$$

and the sample variance is given by:

$$\sigma_a^2 = \frac{1}{N_a}\sum_i (x_{ai} - \mu_a)^2$$

The factor analysis model for this particular sample is then:

$$\begin{array}{ccccccc} z_{1,i} & = & \ell_{1,1}F_{1,i} & + & \ell_{1,2}F_{2,i} & + & \varepsilon_{1,i} \\ \vdots & & \vdots & & \vdots & & \vdots \\ z_{10,i} & = & \ell_{10,1}F_{1,i} & + & \ell_{10,2}F_{2,i} & + & \varepsilon_{10,i} \end{array}$$

or, more succinctly:

$$z_{ai} = \sum_p \ell_{ap}F_{pi} + \varepsilon_{ai}$$

where

- $F_{1,i}$ is the *i*th student's "verbal intelligence",
- $F_{2,i}$ is the *i*th student's "mathematical intelligence",
- ℓ_{ap} are the factor loadings for the *a*th subject, for $p = 1, 2$.

In matrix notation, we have

$$Z = LF + \epsilon$$

Observe that by doubling the scale on which "verbal intelligence"—the first component in each column of *F*—is measured, and simultaneously halving the factor loadings for verbal intelligence makes no difference to the model. Thus, no generality is lost by assuming that the standard deviation of verbal intelligence is 1. Likewise for mathematical intelligence. Moreover, for similar reasons, no generality is lost by assuming the two factors are uncorrelated with each other. In other words:

$$\sum_i F_{pi}F_{qi} = \delta_{pq}$$

where δ_{pq} is the Kronecker delta (0 when $p \neq q$ and 1 when $p = q$).The errors are assumed to be independent of the factors:

$$\sum_i F_{pi}\varepsilon_{ai} = 0$$

Note that, since any rotation of a solution is also a solution, this makes interpreting the factors difficult. In this particular example, if we do not know beforehand that the two types of intelligence are uncorrelated, then we cannot interpret the two factors as the two different types of intelligence. Even if they are uncorrelated, we cannot tell which factor corresponds to verbal intelligence and which corresponds to mathematical intelligence without an outside argument.

The values of the loadings L, the averages μ, and the variances of the "errors" ε must be estimated given the observed data X and F (the assumption about the levels of the factors is fixed for a given F). The "fundamental theorem" may be derived from the above conditions:

$$\sum_i z_{ai} z_{bi} = \sum_p \ell_{ap} \ell_{bp} + \sum_i \varepsilon_{ai} \varepsilon_{bi}$$

The term on the left is just the correlation matrix of the observed data, and its N_a diagonal elements will be 1's. The last term on the right will be a diagonal matrix with terms less than unity. The first term on the right is the "reduced correlation matrix" and will be equal to the correlation matrix except for its diagonal values which will be less than unity. These diagonal elements of the reduced correlation matrix are called "communalities":

$$h_a^2 = 1 - \psi_a = \sum_p \ell_{ap} \ell_{ap}$$

The sample data z_{ai} will not, of course, exactly obey the fundamental equation given above due to sampling errors, inadequacy of the model, etc. The goal of any analysis of the above model is to find the factors F_{pi} and loadings ℓ_{ap} which, in some sense, give a "best fit" to the data. In factor analysis, the best fit is defined as the minimum of the mean square error in the off-diagonal residuals of the correlation matrix:

$$\varepsilon^2 = \sum_{ab, a \neq b} \left[\sum_i z_{ai} z_{bi} - \sum_p \ell_{ap} \ell_{bp} \right]^2$$

This is equivalent to minimizing the off-diagonal components of the error covariance which, in the model equations have expected values of zero. This is to be contrasted with principal component analysis which seeks to minimize the mean square error of all residuals. Before the advent of high speed computers, considerable effort was devoted to finding approximate solutions to the problem, particularly in estimating the communalities by other means, which then simplifies the problem considerably by yielding a known reduced correlation matrix. This was then used to estimate the factors and the loadings. With the advent of high-speed computers, the minimization problem can be solved quickly and directly, and the communalities are calculated in the process, rather than being needed beforehand. The MinRes algorithm is particularly suited to this problem, but is hardly the only means of finding an exact solution.

Geometric Interpretation

The parameters and variables of factor analysis can be given a geometrical interpretation. The data (z_{ai}), the factors (F_{pi}) and the errors (ε_{ai}) can be viewed as vectors in an N_i-dimensional Euclidean space (sample space), represented as $\mathbf{z}_a$, $\mathbf{F}_p$ and $\boldsymbol{\varepsilon}_a$ respectively. Since the data is standardized, the data vectors are of unit length ($\mathbf{z}_a \cdot \mathbf{z}_a = 1$). The factor vectors define an -dimensional linear subspace (i.e. a hyperplane) in this space, upon which the data vectors are projected orthogonally. This follows from the model equation

$$\mathbf{z}_a = \sum_p \ell_{ap} \mathbf{F}_p + \boldsymbol{\varepsilon}_a$$

and the independence of the factors and the errors: $\mathbf{F}_p \cdot \boldsymbol{\varepsilon}_a = 0$. In the above example, the hyperplane is just a 2-dimensional plane defined by the two factor vectors. The projection of the data vectors onto the hyperplane is given by

$$\hat{\mathbf{z}}_a = \sum_p \ell_{ap} \mathbf{F}_p$$

and the errors are vectors from that projected point to the data point and are perpendicular to the hyperplane. The goal of factor analysis is to find a hyperplane which is a "best fit" to the data in some sense, so it doesn't matter how the factor vectors which define this hyperplane are chosen, as long as they are independent and lie in the hyperplane. We are free to specify them as both orthogonal and normal ($\mathbf{F}_p \cdot \mathbf{F}_q = \delta_{pq}$) with no loss of generality. After a suitable set of factors are found, they may also be arbitrarily rotated within the hyperplane, so that any rotation of the factor vectors will define the same hyperplane, and also be a solution.

As a result, in the above example, in which the fitting hyperplane is two dimensional, if we do not know beforehand that the two types of intelligence are uncorrelated, then we cannot interpret the two factors as the two different types of intelligence. Even if they are uncorrelated, we cannot tell which factor corresponds to verbal intelligence and which corresponds to mathematical intelligence, or whether the factors are linear combinations of both, without an outside argument.

The data vectors $\mathbf{z}_a$ have unit length. The correlation matrix for the data is given by $r_{ab} = \mathbf{z}_a \cdot \mathbf{z}_b$. The correlation matrix can be geometrically interpreted as the cosine of the angle between the two data vectors $\mathbf{z}_a$ and $\mathbf{z}_b$. The diagonal elements will clearly be 1's and the off diagonal elements will have absolute values less than or equal to unity. The "reduced correlation matrix" is defined as

$$\hat{r}_{ab} = \hat{\mathbf{z}}_a \cdot \hat{\mathbf{z}}_b .$$

The goal of factor analysis is to choose the fitting hyperplane such that the reduced correlation matrix reproduces the correlation matrix as nearly as possible, except for the diagonal elements of the correlation matrix which are known to have unit value. In other words, the goal is to reproduce as accurately as possible the cross-correlations in the data. Specifically, for the fitting hyperplane, the mean square error in the off-diagonal components

$$\varepsilon^2 = \sum_{a,b \neq a} \left(r_{ab} - \hat{r}_{ab} \right)^2$$

is to be minimized, and this is accomplished by minimizing it with respect to a set of orthonormal factor vectors. It can be seen that

$$r_{ab} - \hat{r}_{ab} = \boldsymbol{\varepsilon}_a \cdot \boldsymbol{\varepsilon}_b$$

The term on the right is just the covariance of the errors. In the model, the error covariance is stated to be a diagonal matrix and so the above minimization problem will in fact yield a "best fit" to the model: It will yield a sample estimate of the error covariance which has its off-diagonal components minimized in the mean square sense. It can be seen that since the $\hat{z}_a$ are orthogonal projections of the data vectors, their length will be less than or equal to the length of the projected data vector, which is unity. The square of these lengths are just the diagonal elements of the reduced correlation matrix. These diagonal elements of the reduced correlation matrix are known as "communalities":

$$h_a^2 = \hat{\mathbf{z}}_a \cdot \hat{\mathbf{z}}_a = \sum_p \ell_{ap} \ell_{ap}$$

Large values of the communalities will indicate that the fitting hyperplane is rather accurately reproducing the correlation matrix. It should be noted that the mean values of the factors must also be constrained to be zero, from which it follows that the mean values of the errors will also be zero.

Practical Implementation

Type of Factor Analysis

Exploratory factor analysis (EFA) is used to identify complex interrelationships among items and group items that are part of unified concepts. The researcher makes no "a priori" assumptions about relationships among factors.

Confirmatory factor analysis (CFA) is a more complex approach that tests the hypothesis that the items are associated with specific factors. CFA uses structural equation modelling to test a measurement model whereby loading on the factors allows for evaluation of relationships between observed variables and unobserved variables. Structural equation modelling approaches can accommodate measurement error, and are less restrictive than least-squares estimation. Hypothesized models are tested against actual data, and the analysis would demonstrate loadings of observed variables on the latent variables (factors), as well as the correlation between the latent variables.

Types of Factoring

Principal component analysis (PCA): PCA is a widely used method for factor extraction, which is the first phase of EFA. Factor weights are computed in order to extract the maximum possible variance, with successive factoring continuing until there is no further meaningful variance left. The factor model must then be rotated for analysis.

Canonical factor analysis, also called Rao's canonical factoring, is a different method of computing the same model as PCA, which uses the principal axis method. Canonical factor analysis seeks factors which have the highest canonical correlation with the observed variables. Canonical factor analysis is unaffected by arbitrary rescaling of the data.

Common factor analysis, also called principal factor analysis (PFA) or principal axis factoring (PAF), seeks the least number of factors which can account for the common variance (correlation) of a set of variables.

Image factoring: based on the correlation matrix of predicted variables rather than actual variables, where each variable is predicted from the others using multiple regression.

Alpha factoring: based on maximizing the reliability of factors, assuming variables are randomly sampled from a universe of variables. All other methods assume cases to be sampled and variables fixed.

Factor regression model: a combinatorial model of factor model and regression model; or alternatively, it can be viewed as the hybrid factor model, whose factors are partially known.

Terminology

Factor loadings: The factor loadings, also called component loadings in PCA (not sure in Factor Analysis), are the correlation coefficients between the cases (rows) and factors (columns). Analogous to Pearson's r, the squared factor loading is the percent of variance in that indicator variable explained by the factor. To get the percent of variance in all the variables accounted for by each factor, add the sum of the squared factor loadings for that factor (column) and divide by the number of variables. (Note the number of variables equals the sum of their variances as the variance of a standardized variable is 1.) This is the same as dividing the factor's eigenvalue by the number of variables.

Interpreting factor loadings: By one rule of thumb in confirmatory factor analysis, loadings should be .7 or higher to confirm that independent variables identified a priori are represented by a particular factor, on the rationale that the .7 level corresponds to about half of the variance in the indicator being explained by the factor. However, the .7 standard is a high one and real-life data may well not meet this criterion, which is why some researchers, particularly for exploratory purposes, will use a lower level such as .4 for the central factor and .25 for other factors. In any event, factor loadings must be interpreted in the light of theory, not by arbitrary cutoff levels.

In oblique rotation, one gets both a pattern matrix and a structure matrix. The structure matrix is simply the factor loading matrix as in orthogonal rotation, representing the variance in a measured variable explained by a factor on both a unique and common contributions basis. The pattern matrix, in contrast, contains coefficients which just represent unique contributions. The more factors, the lower the pattern coefficients as a rule since there will be more common contributions to variance explained. For oblique rotation, the researcher looks at both the structure and pattern coefficients when attributing a label to a factor. Principles of oblique rotation can be derived from both cross entropy and its dual entropy.

Communality: The sum of the squared factor loadings for all factors for a given variable (row) is the variance in that variable accounted for by all the factors, and this is called the communality. The communality measures the percent of variance in a given variable explained by all the factors jointly and may be interpreted as the reliability of the indicator.

Spurious solutions: If the communality exceeds 1.0, there is a spurious solution, which may reflect too small a sample or the researcher has too many or too few factors.

Uniqueness of a variable: That is, uniqueness is the variability of a variable minus its communality.

Eigenvalues:/Characteristic roots: The eigenvalue for a given factor measures the variance in all the variables which is accounted for by that factor. The ratio of eigenvalues is the ratio of explanatory importance of the factors with respect to the variables. If a factor has a low eigenvalue, then it is contributing little to the explanation of variances in the variables and may be ignored as redundant with more important factors. Eigenvalues measure the amount of variation in the total sample accounted for by each factor.

Extraction sums of squared loadings: Initial eigenvalues and eigenvalues after extraction (listed by SPSS as "Extraction Sums of Squared Loadings") are the same for PCA extraction, but for other extraction methods, eigenvalues after extraction will be lower than their initial counterparts. SPSS also prints "Rotation Sums of Squared Loadings" and even for PCA, these eigenvalues will differ from initial and extraction eigenvalues, though their total will be the same.

Factor scores (also called component scores in PCA): are the scores of each case (row) on each factor (column). To compute the factor score for a given case for a given factor, one takes the case's standardized score on each variable, multiplies by the corresponding loadings of the variable for the given factor, and sums these products. Computing factor scores allows one to look for factor outliers. Also, factor scores may be used as variables in subsequent modelling. (Explained from PCA not from Factor Analysis perspective).

Criteria for Determining the Number of Factors

Using one or more of the methods below, the researcher determines an appropriate range of solutions to investigate. Methods may not agree. For instance, the Kaiser criterion may suggest five factors and the scree test may suggest two, so the researcher may request 3-, 4-, and 5-factor solutions discuss each in terms of their relation to external data and theory.

Comprehensibility: A purely subjective criterion would be to retain those factors whose meaning is comprehensible to the researcher. This is not recommended.

Kaiser criterion: The Kaiser rule is to drop all components with eigenvalues under 1.0 – this being the eigenvalue equal to the information accounted for by an average single item. The Kaiser criterion is the default in SPSS and most statistical software but is not recommended when used as the sole cut-off criterion for estimating the number of factors as it tends to overextract factors. A variation of this method has been created where a researcher calculates confidence intervals for each eigenvalue and retains only factors which have the entire confidence interval greater than 1.0.

Variance explained criteria: Some researchers simply use the rule of keeping enough factors to account for 90% (sometimes 80%) of the variation. Where the researcher's goal emphasizes parsimony (explaining variance with as few factors as possible), the criterion could be as low as 50%

Scree plot: The Cattell scree test plots the components as the X axis and the corresponding eigenvalues as the Y-axis. As one moves to the right, toward later components, the eigenvalues drop. When the drop ceases and the curve makes an elbow toward less steep decline, Cattell's scree test says to drop all further components after the one starting the elbow. This rule is sometimes criticised for being amenable to researcher-controlled "fudging". That is, as picking the "elbow" can be subjective because the curve has multiple elbows or is a smooth curve, the researcher may be tempted to set the cut-off at the number of factors desired by his or her research agenda.

Horn's Parallel Analysis (PA): A Monte-Carlo based simulation method that compares the observed eigenvalues with those obtained from uncorrelated normal variables. A factor or component is retained if the associated eigenvalue is bigger than the 95th of the distribution of eigenvalues derived from the random data. PA is one of the most recommendable rules for determining the number of components to retain, but only few programs include this option.

However, before dropping a factor below one's cutoff, the analyst(s) should create a data set based on the factor loadings and check the scores' correlation with any given dependent variable(s) of interest. Scores based on a factor with a very small eigenvalue can correlate strongly with dependent variables, in which case dropping such a factor from a theoretical model may reduce its predictive validity.

Velicer's (1976) MAP test "involves a complete principal components analysis followed by the examination of a series of matrices of partial correlations" (p. 397). The squared correlation for Step "0" is the average squared off-diagonal correlation for the unpartialed correlation matrix. On Step 1, the first principal component and its associated items are partialed out. Thereafter, the average squared off-diagonal correlation for the subsequent correlation matrix is then computed for Step 1. On Step 2, the first two principal components are partialed out and the resultant average squared off-diagonal correlation is again computed. The computations are carried out for k minus one step (k representing the total number of variables in the matrix). Thereafter, all of the average squared correlations for each step are lined up and the step number in the analyses that resulted in the lowest average squared partial correlation determines the number of components or factors to retain (Velicer, 1976). By this method, components are maintained as long as the variance in the correlation matrix represents systematic variance, as opposed to residual or error variance. Although methodologically akin to principal components analysis, the MAP technique has been shown to perform quite well in determining the number of factors to retain in multiple simulation studies. This procedure is made available through SPSS's user interface.

Rotation Methods

The unrotated output maximises variance accounted for by the first and subsequent factors, and forcing the factors to be orthogonal. This data-compression comes at the cost of having most items load on the early factors, and usually, of having many items load substantially on more than one factor. Rotation serves to make the output more understandable, by seeking so-called "Simple Structure": A pattern of loadings where items load most strongly on one factor, and much more weakly on the other factors. Rotations can be orthogonal or oblique (allowing the factors to correlate).

Varimax rotation is an orthogonal rotation of the factor axes to maximize the variance of the squared loadings of a factor (column) on all the variables (rows) in a factor matrix, which has the effect of differentiating the original variables by extracted factor. Each factor will tend to have either large or small loadings of any particular variable. A varimax solution yields results which make it as easy as possible to identify each variable with a single factor. This is the most common rotation option. However, the orthogonality (i.e., independence) of factors is often an unrealistic assumption. Oblique rotations are inclusive of orthogonal rotation, and for that reason, oblique rotations are a preferred method.

Quartimax rotation is an orthogonal alternative which minimizes the number of factors needed to explain each variable. This type of rotation often generates a general factor on which most variables are loaded to a high or medium degree. Such a factor structure is usually not helpful to the research purpose.

Equimax rotation is a compromise between Varimax and Quartimax criteria.

Direct oblimin rotation is the standard method when one wishes a non-orthogonal (oblique) solution – that is, one in which the factors are allowed to be correlated. This will result in higher eigenvalues but diminished interpretability of the factors.

Promax rotation is an alternative non-orthogonal (oblique) rotation method which is computationally faster than the direct oblimin method and therefore is sometimes used for very large datasets.

Factor Analysis in Psychometrics

History

Charles Spearman pioneered the use of factor analysis in the field of psychology and is sometimes credited with the invention of factor analysis. He discovered that school children's scores on a wide variety of seemingly unrelated subjects were positively correlated, which led him to postulate that a general mental ability, or *g*, underlies and shapes human cognitive performance. His postulate now enjoys broad support in the field of intelligence research, where it is known as the *g* theory.

Raymond Cattell expanded on Spearman's idea of a two-factor theory of intelligence after performing his own tests and factor analysis. He used a multi-factor theory to explain intelligence. Cattell's theory addressed alternate factors in intellectual development, including motivation and psychology. Cattell also developed several mathematical methods for adjusting psychometric graphs, such as his "scree" test and similarity coefficients. His research led to the development of his theory of fluid and crystallized intelligence, as well as his 16 Personality Factors theory of personality. Cattell was a strong advocate of factor analysis and psychometrics. He believed that all theory should be derived from research, which supports the continued use of empirical observation and objective testing to study human intelligence.

Applications in Psychology

Factor analysis is used to identify "factors" that explain a variety of results on different tests. For example, intelligence research found that people who get a high score on a test of verbal ability are also good on other tests that require verbal abilities. Researchers explained this by using factor analysis to isolate one factor, often called crystallized intelligence or verbal intelligence, which represents the degree to which someone is able to solve problems involving verbal skills.

Factor analysis in psychology is most often associated with intelligence research. However, it also has been used to find factors in a broad range of domains such as personality, attitudes, beliefs, etc. It is linked to psychometrics, as it can assess the validity of an instrument by finding if the instrument indeed measures the postulated factors.

Advantages

- Reduction of number of variables, by combining two or more variables into a single factor. For example, performance at running, ball throwing, batting, jumping and weight lifting could be combined into a single factor such as general athletic ability. Usually, in an item by people matrix, factors are selected by grouping related items. In the Q factor analysis technique, the matrix is transposed and factors are created by grouping related people: For example, liberals, libertarians, conservatives and socialists, could form separate groups.
- Identification of groups of inter-related variables, to see how they are related to each other. For example, Carroll used factor analysis to build his Three Stratum Theory. He found that a factor called "broad visual perception" relates to how good an individual is at visual tasks. He also found a "broad auditory perception" factor, relating to auditory task capability. Furthermore, he found a global factor, called "g" or general intelligence, that relates to both "broad visual perception" and "broad auditory perception". This means someone with a high "g" is likely to have both a high "visual perception" capability and a high "auditory perception" capability, and that "g" therefore explains a good part of why someone is good or bad in both of those domains.

Disadvantages

- "...each orientation is equally acceptable mathematically. But different factorial theories proved to differ as much in terms of the orientations of factorial axes for a given solution as in terms of anything else, so that model fitting did not prove to be useful in distinguishing among theories." (Sternberg, 1977). This means all rotations represent different underlying processes, but all rotations are equally valid outcomes of standard factor analysis optimization. Therefore, it is impossible to pick the proper rotation using factor analysis alone.

- Factor analysis can be only as good as the data allows. In psychology, where researchers often have to rely on less valid and reliable measures such as self-reports, this can be problematic.
- Interpreting factor analysis is based on using a "heuristic", which is a solution that is "convenient even if not absolutely true". More than one interpretation can be made of the same data factored the same way, and factor analysis cannot identify causality.

Exploratory Factor Analysis versus Principal Components Analysis

While exploratory factor analysis and principal component analysis are treated as synonymous techniques in some fields of statistics, this has been criticised (e.g. Fabrigar et al., 1999; Suhr, 2009). In factor analysis, the researcher makes the assumption that an underlying causal model exists, whereas PCA is simply a variable reduction technique. Researchers have argued that the distinctions between the two techniques may mean that there are objective benefits for preferring one over the other based on the analytic goal. If the factor model is incorrectly formulated or the assumptions are not met, then factor analysis will give erroneous results. Factor analysis has been used successfully where adequate understanding of the system permits good initial model formulations. Principal component analysis employs a mathematical transformation to the original data with no assumptions about the form of the covariance matrix. The aim of PCA is to determine a few linear combinations of the original variables that can be used to summarize the data set without losing much information.

Arguments Contrasting PCA and EFA

Fabrigar et al. (1999) address a number of reasons used to suggest that principal components analysis is not equivalent to factor analysis:

1. It is sometimes suggested that principal components analysis is computationally quicker and requires fewer resources than factor analysis. Fabrigar et al. suggest that the ready availability of computer resources have rendered this practical concern irrelevant.
2. PCA and factor analysis can produce similar results. This point is also addressed by Fabrigar et al.; in certain cases, whereby the communalities are low (e.g., .40), the two techniques produce divergent results. In fact, Fabrigar et al. argue that in cases where the data correspond to assumptions of the common factor model, the results of PCA are inaccurate results.

3. There are certain cases where factor analysis leads to 'Heywood cases'. These encompass situations whereby 100% or more of the variance in a measured variable is estimated to be accounted for by the model. Fabrigar et al. suggest that these cases are actually informative to the researcher, indicating a misspecified model or a violation of the common factor model. The lack of Heywood cases in the PCA approach may mean that such issues pass unnoticed.
4. Researchers gain extra information from a PCA approach, such as an individual's score on a certain component – such information is not yielded from factor analysis. However, as Fabrigar et al. contend, the typical aim of factor analysis – i.e. to determine the factors accounting for the structure of the correlations between measured variables – does not require knowledge of factor scores and thus this advantage is negated. It is also possible to compute factor scores from a factor analysis.

Variance versus Covariance

Factor analysis takes into account the random error that is inherent in measurement, whereas PCA fails to do so. This point is exemplified by Brown (2009), who indicated that, in respect to the correlation matrices involved in the calculations:

> *"In PCA, 1.00s are put in the diagonal meaning that all of the variance in the matrix is to be accounted for (including variance unique to each variable, variance common among variables, and error variance). That would, therefore, by definition, include all of the variance in the variables. In contrast, in EFA, the communalities are put in the diagonal meaning that only the variance shared with other variables is to be accounted for (excluding variance unique to each variable and error variance). That would, therefore, by definition, include only variance that is common among the variables."*
> — *Brown (2009),* Principal components analysis and exploratory factor analysis – Definitions, differences and choices

For this reason, Brown (2009) recommends using factor analysis when theoretical ideas about relationships between variables exist, whereas PCA should be used if the goal of the researcher is to explore patterns in their data.

Differences in Procedure and Results

The differences between principal components analysis and factor analysis are further illustrated by Suhr (2009):

- PCA results in principal components that account for a maximal amount of variance for observed variables; FA account for common variance in the data.
- PCA inserts ones on the diagonals of the correlation matrix; FA adjusts the diagonals of the correlation matrix with the unique factors.
- PCA minimizes the sum of squared perpendicular distance to the component axis; FA estimates factors which influence responses on observed variables.
- The component scores in PCA represent a linear combination of the observed variables weighted by eigenvectors; the observed variables in FA are linear combinations of the underlying and unique factors.
- In PCA, the components yielded are uninterpretable, i.e. they do not represent underlying 'constructs'; in FA, the underlying constructs can be labelled and readily interpreted, given an accurate model specification.

Factor Analysis in Marketing

The basic steps are:

- Identify the salient attributes consumers use to evaluate products in this category.
- Use quantitative marketing research techniques (such as surveys) to collect data from a sample of potential customers concerning their ratings of all the product attributes.
- Input the data into a statistical program and run the factor analysis procedure. The computer will yield a set of underlying attributes (or factors).
- Use these factors to construct perceptual maps and other product positioning devices.

Information Collection

The data collection stage is usually done by marketing research professionals. Survey questions ask the respondent to rate a product sample or descriptions of product concepts on a range of attributes. Anywhere from five to twenty attributes are chosen. They could include things like: ease of use, weight, accuracy, durability, colourfulness, price, or size. The attributes chosen will vary depending on the product being studied. The same question is asked about all the products in the study. The data for multiple products is coded and input into a statistical program such as R, SPSS, SAS, Stata, STATISTICA, JMP, and SYSTAT.

Analysis

The analysis will isolate the underlying factors that explain the data using a matrix of associations. Factor analysis is an interdependence technique. The complete set of interdependent relationships is examined. There is no specification of dependent variables, independent variables, or causality. Factor analysis assumes that all the rating data on different attributes can be reduced down to a few important dimensions. This reduction is possible because some attributes may be related to each other. The rating given to any one attribute is partially the result of the influence of other attributes. The statistical algorithm deconstructs the rating (called a raw score) into its various components, and reconstructs the partial scores into underlying factor scores. The degree of correlation between the initial raw score and the final factor score is called a *factor loading.*

Advantages

- Both objective and subjective attributes can be used provided the subjective attributes can be converted into scores.
- Factor analysis can identify latent dimensions or constructs that direct analysis may not.
- It is easy and inexpensive.

Disadvantages

- Usefulness depends on the researchers' ability to collect a sufficient set of product attributes. If important attributes are excluded or neglected, the value of the procedure is reduced.
- If sets of observed variables are highly similar to each other and distinct from other items, factor analysis will assign a single factor to them. This may obscure factors that represent more interesting relationships.
- Naming factors may require knowledge of theory because seemingly dissimilar attributes can correlate strongly for unknown reasons.

Factor Analysis in Physical and Biological Sciences

Factor analysis has also been widely used in physical sciences such as geochemistry, ecology, hydrochemistry., astrophysics, cosmology, as well as biological sciences such as molecular biology and biochemistry. In groundwater quality management, it is important to relate the spatial distribution of different chemical parameters to different possible sources, which have different chemical signatures. For example, a sulfide mine is likely to be associated with high levels of acidity, dissolved sulfates and transition metals. These signatures can be identified as factors through R-mode factor analysis, and the location of possible sources can be suggested by contouring the factor scores.

In geochemistry, different factors can correspond to different mineral associations, and thus to mineralisation.

Factor Analysis in Microarray Analysis

Factor analysis can be used for summarizing high-density oligonucleotide DNA microarrays data at probe level for Affymetrix GeneChips. In this case, the latent variable corresponds to the RNA concentration in a sample.

Implementation

Factor analysis has been implemented in several statistical analysis programs since the 1980s: SAS, BMDP and SPSS. It is also implemented in the *R* programming language (with the *factanal* function), OpenOpt, and the statistical software package Stata. Rotations are implemented in the *GPArotation* R package.

Correlation and Dependence

In statistics, dependence is any statistical relationship between two random variables or two sets of data. Correlation refers to any of a broad class of statistical relationships involving dependence.

Familiar examples of dependent phenomena include the correlation between the physical statures of parents and their offspring, and the correlation between the demand for a product and its price. Correlations are useful because they can indicate a predictive relationship that can be exploited in practice. For example, an electrical utility may produce less power on a mild day based on the correlation between electricity demand and weather. In this example there is a causal relationship, because extreme weather causes people to use more electricity for heating or cooling; however, statistical dependence is not sufficient to demonstrate the presence of such a causal relationship (i.e., correlation does not imply causation).

Formally, *dependence* refers to any situation in which random variables do not satisfy a mathematical condition of probabilistic independence. In loose usage, *correlation* can refer to any departure of two or more random variables from independence, but technically it refers to any of several more specialised types of relationship between mean values. There are several correlation coefficients, often denoted ρ or *r*, measuring the degree of correlation. The most common of these is the Pearson correlation coefficient, which is sensitive only to a linear relationship between two variables (which may exist even if one is a nonlinear function of the other). Other correlation coefficients have been developed to be more robust than the Pearson correlation – that is, more sensitive to nonlinear relationships. Mutual information can also be applied to measure dependence between two variables.

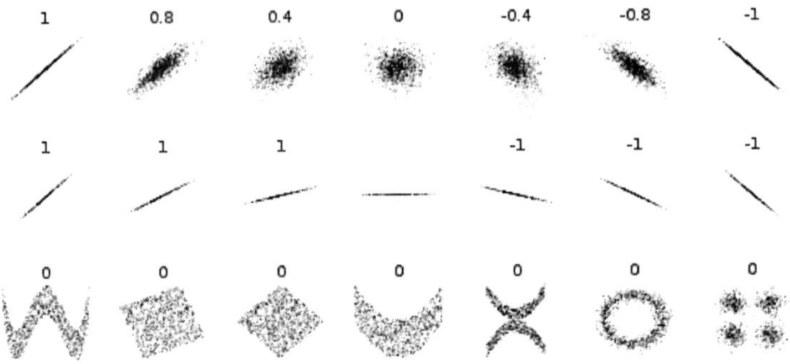

Figure: *Several sets of* (x, y) *points, with the Pearson correlation coefficient of* x *and* y *for each set. Note that the correlation reflects the noisiness and direction of a linear relationship (top row), but not the slope of that relationship (middle), nor many aspects of nonlinear relationships (bottom). N.B.: the figure in the centre has a slope of 0 but in that case the correlation coefficient is undefined because the variance of* Y *is zero.*

Pearson's Product-Moment Coefficient

The most familiar measure of dependence between two quantities is the Pearson product-moment correlation coefficient, or "Pearson's correlation coefficient", commonly called simply "the correlation coefficient". It is obtained by dividing the covariance of the two variables by the product of their standard deviations. Karl Pearson developed the coefficient from a similar but slightly different idea by Francis Galton.

The population correlation coefficient $\rho_{X,Y}$ between two random variables X and Y with expected values μ_X and μ_Y and standard deviations σ_X and σ_Y is defined as:

$$\rho_{X,Y} = \mathrm{corr}(X,Y) = \frac{\mathrm{cov}(X,Y)}{\sigma_X \sigma_Y} = \frac{E[(X-\mu_X)(Y-\mu_Y)]}{\sigma_X \sigma_Y},$$

where E is the expected value operator, *cov* means covariance, and, *corr* a widely used alternative notation for the correlation coefficient.

The Pearson correlation is defined only if both of the standard deviations are finite and nonzero. It is a corollary of the Cauchy–Schwarz inequality that the correlation cannot exceed 1 in absolute value. The correlation coefficient is symmetric: corr(X, Y) = corr(Y,X).

The Pearson correlation is +1 in the case of a perfect direct (increasing) linear relationship (correlation), –1 in the case of a perfect decreasing (inverse) linear relationship (anticorrelation), and some value between –1 and 1 in all other cases, indicating the degree of linear dependence between the variables. As it approaches zero there is less of a relationship (closer to uncorrelated). The closer the coefficient is to either "1 or 1, the stronger the correlation between the variables.

If the variables are independent, Pearson's correlation coefficient is 0, but the converse is not true because the correlation coefficient detects only linear dependencies between two variables. For example, suppose the random variable X is symmetrically distributed about zero, and $Y = X^2$. Then Y is completely determined by X, so that X and Y are perfectly dependent, but their correlation is zero; they are uncorrelated. However, in the special case when X and Y are jointly normal, uncorrelatedness is equivalent to independence.

If we have a series of n measurements of X and Y written as x_i and y_i where i – 1, 2, ..., n, then the *sample correlation coefficient* can be used to estimate the population Pearson correlation r between X and Y. The sample correlation coefficient is written

$$r_{xy} = \frac{\sum_{i=1}^{n}(x_i - \bar{x})(y_i - \bar{y})}{(n-1)s_x s_y} = \frac{\sum_{i=1}^{n}(x_i - \bar{x})(y_i - \bar{y})}{\sqrt{\sum_{i=1}^{n}(x_i - \bar{x})^2 \sum_{i=1}^{n}(y_i - \bar{y})^2}},$$

where x and y are the sample means of X and Y, and s_x and s_y are the sample standard deviations of X and Y.

This can also be written as:

$$r_{xy} = \frac{\sum x_i y_i - n\overline{xy}}{(n-1)s_x s_y} = \frac{n\sum x_i y_i - \sum x_i \sum y_i}{\sqrt{n\sum x_i^2 - (\sum x_i)^2}\sqrt{n\sum y_i^2 - (\sum y_i)^2}}.$$

If x and y are results of measurements that contain measurement error, the realistic limits on the correlation coefficient are not –1 to +1 but a smaller range.

For the case of a linear model with a single independent variable, the coefficient of determination (R squared) is the square of r, Pearson's product-moment coefficient .

Rank Correlation Coefficients

Rank correlation coefficients, such as Spearman's rank correlation coefficient and Kendall's rank correlation coefficient (τ) measure the extent to which, as one variable increases, the other variable tends to increase, without requiring that increase to be represented by a linear relationship. If, as the one variable increases, the other *decreases*, the rank correlation coefficients will be negative. It is common to regard these rank correlation coefficients as alternatives to Pearson's coefficient, used either to reduce the amount of calculation or to make the coefficient less sensitive to non-normality in distributions. However, this view has little mathematical basis, as rank correlation coefficients measure a different type of relationship than the Pearson product-moment correlation coefficient, and are best seen as measures of a different type of association, rather than as alternative measure of the population correlation coefficient.

To illustrate the nature of rank correlation, and its difference from linear correlation, consider the following four pairs of numbers (x, y):

$$(0, 1), (10, 100), (101, 500), (102, 2000).$$

As we go from each pair to the next pair x increases, and so does y. This relationship is perfect, in the sense that an increase in x is *always* accompanied by an increase in y. This means that we have a perfect rank correlation, and both Spearman's and Kendall's correlation coefficients are 1, whereas in this example Pearson product-moment correlation coefficient is 0.7544, indicating that the points are far from lying on a straight line. In the same way if y always *decreases* when x *increases*, the rank correlation coefficients will be –1, while the Pearson product-moment correlation coefficient may or may not be close to –1, depending on how close the points are to a straight line. Although in the extreme cases of perfect rank correlation the two coefficients are both equal (being both +1 or both –1) this is not in general so, and values of the two coefficients cannot meaningfully be compared. For example, for the three pairs (1, 1) (2, 3) (3, 2) Spearman's coefficient is 1/2, while Kendall's coefficient is 1/3.

Other Measures of Dependence among Random Variables

The information given by a correlation coefficient is not enough to define the dependence structure between random variables. The correlation coefficient completely defines the dependence structure only in very particular cases, for example when the distribution is a multivariate normal distribution. In the case of elliptical distributions it characterizes the (hyper-)ellipses of equal density, however, it does not completely characterize the dependence structure (for example, a multivariate t-distribution's degrees of freedom determine the level of tail dependence).

Distance correlation and Brownian covariance / Brownian correlation were introduced to address the deficiency of Pearson's correlation that it can be zero for dependent random variables; zero distance correlation and zero Brownian correlation imply independence.

The Randomized Dependence Coefficient is a computationally efficient, copula-based measure of dependence between multivariate random variables. RDC is invariant with respect to non-linear scalings of random variables, is capable of discovering a wide range of functional association patterns and takes value zero at independence.

The correlation ratio is able to detect almost any functional dependency, and the entropy-based mutual information, total correlation and dual total correlation are capable of detecting even more general dependencies. These are sometimes referred to as multi-moment correlation measures, in comparison to those that consider only second moment (pairwise or quadratic) dependence.

The polychoric correlation is another correlation applied to ordinal data that aims to estimate the correlation between theorised latent variables.

One way to capture a more complete view of dependence structure is to consider a copula between them.

The coefficient of determination generalizes the correlation coefficient for relationships beyond simple linear regression.

Sensitivity to the Data Distribution

The degree of dependence between variables X and Y does not depend on the scale on which the variables are expressed. That is, if we are analyzing the relationship between X and Y, most correlation measures are unaffected by transforming X to $a + bX$ and Y to $c + dY$, where a, b, c, and d are constants. This is true of some correlation statistics as well as their population analogues. Some correlation statistics, such as the rank correlation coefficient, are also invariant to monotone transformations of the marginal distributions of X and/or Y.

Most correlation measures are sensitive to the manner in which X and Y are sampled. Dependencies tend to be stronger if viewed over a wider range of values. Thus, if we consider the correlation coefficient between the heights of fathers and their sons over all adult males, and compare it to the same correlation coefficient calculated when the fathers are selected to be between 165 cm and 170 cm in height, the correlation will be weaker in the latter case. Several techniques have been developed that attempt to correct for range restriction in one or both variables, and are commonly used in meta-analysis; the most common are Thorndike's case II and case III equations.

Various correlation measures in use may be undefined for certain joint distributions of X and Y. For example, the Pearson correlation coefficient is defined in terms of moments, and hence will be undefined if the moments are undefined. Measures of dependence based on quantiles are always defined. Sample-based statistics intended to estimate population measures of dependence may or may not have desirable statistical properties such as being unbiased, or asymptotically consistent, based on the spatial structure of the population from which the data were sampled.

Sensitivity to the data distribution can be used to an advantage. For example, scaled correlation is designed to use the sensitivity to the range in order to pick out correlations between fast components of time series. By reducing the range of values in a controlled manner, the correlations on long time scale are filtered out and only the correlations on short time scales are revealed.

Correlation Matrices

The correlation matrix of n random variables $X_1, ..., X_n$ is the $n \times n$ matrix whose i,j entry is $\text{corr}(X_i, X_j)$. If the measures of correlation used are product-moment coefficients, the correlation matrix is the same as the covariance matrix of the standardized random variables $X_i / \sigma(X_i)$ for $i = 1, ..., n$. This applies to both the matrix of population correlations (in which case "σ" is the population standard deviation), and to the matrix of sample correlations (in which case "σ" denotes the sample standard deviation). Consequently, each is necessarily a positive-semidefinite matrix.

The correlation matrix is symmetric because the correlation between X_i and X_j is the same as the correlation between X_j and X_i.

Common Misconceptions

Correlation and Causality

The conventional dictum that "correlation does not imply causation" means that correlation cannot be used to infer a causal relationship between the variables. This dictum should not be taken to mean that correlations cannot indicate the potential existence of causal relations. However, the causes underlying the correlation, if any, may be indirect and unknown, and high correlations also overlap with identity relations (tautologies), where no causal process exists. Consequently, establishing a correlation between two variables is not a sufficient condition to establish a causal relationship (in either direction).

A correlation between age and height in children is fairly causally transparent, but a correlation between mood and health in people is less so. Does improved mood lead to improved health, or does good health lead to good mood, or both? Or does some other factor underlie both? In other words, a correlation can be taken as evidence for a possible causal relationship, but cannot indicate what the causal relationship, if any, might be.

Correlation and Linearity

The Pearson correlation coefficient indicates the strength of a linear relationship between two variables, but its value generally does not completely characterize their relationship . In particular, if the conditional mean of Y given X, denoted $E(Y|X)$, is not linear in X, the correlation coefficient will not fully determine the form of $E(Y|X)$.

The image on the right shows scatterplots of Anscombe's quartet, a set of four different pairs of variables created by Francis Anscombe. The four y variables have the same mean (7.5), variance (4.12), correlation (0.816) and regression line ($y = 3 + 0.5x$). However, as can be seen on the plots, the distribution of the variables is very different. The first one (top left) seems to be distributed normally, and corresponds to what one would expect when considering two variables correlated and following the assumption of normality. The second one (top right) is not distributed normally; while an obvious relationship between the two variables can be observed, it is not linear. In this case the Pearson correlation coefficient does not indicate that there is an exact functional relationship: only the extent to which that relationship can be approximated by a linear relationship. In the third case (bottom left), the linear relationship is perfect, except for one outlier which exerts enough influence to lower the correlation coefficient from 1 to 0.816. Finally, the fourth example (bottom right) shows another example when one outlier is enough to produce a high correlation coefficient, even though the relationship between the two variables is not linear.

These examples indicate that the correlation coefficient, as a summary statistic, cannot replace visual examination of the data. Note that the examples are sometimes said to demonstrate that the Pearson correlation assumes that the data follow a normal distribution, but this is not correct.

Bivariate Normal Distribution

If a pair (X, Y) of random variables follows a bivariate normal distribution, the conditional mean $E(X \mid Y)$ is a linear function of Y, and the conditional mean $E(Y \mid X)$ is a linear function of X. The correlation coefficient r between X and Y, along with the marginal means and variances of X and Y, determines this linear relationship:

$$E(Y \mid X) = E(Y) + r\sigma_y \frac{X - E(X)}{\sigma_x},$$

where $E(X)$ and $E(Y)$ are the expected values of X and Y, respectively, and σ_x and σ_y are the standard deviations of X and Y, respectively.

Partial Correlation

If a population or data-set is characterized by more than two variables, a partial correlation coefficient measures the strength of dependence between a pair of variables that is not accounted for by the way in which they both change in response to variations in a selected subset of the other variables.

Regression Analysis

In statistics, regression analysis is a statistical process for estimating the relationships among variables. It includes many techniques for modelling and analyzing several variables, when the focus is on the relationship between a dependent variable and one or more independent variables. More specifically, regression analysis helps one understand how the typical value of the dependent variable (or 'criterion variable') changes when any one of the independent variables is varied, while the other independent variables are held fixed. Most commonly, regression analysis estimates the conditional expectation of the dependent variable given the independent variables – that is, the average value of the dependent variable when the independent variables are fixed. Less commonly, the focus is on a quantile, or other location parameter of the conditional distribution of the dependent variable given the independent variables. In all cases, the estimation target is a function of the independent variables called the regression function. In regression analysis, it is also of interest to characterize the variation of the dependent variable around the regression function which can be described by a probability distribution.

Regression analysis is widely used for prediction and forecasting, where its use has substantial overlap with the field of machine learning. Regression analysis is also used to understand which among the independent variables are related to the dependent variable, and to explore the forms of these relationships. In restricted circumstances, regression analysis can be used to infer causal relationships between the independent and dependent variables. However this can lead to illusions or false relationships, so caution is advisable; for example, correlation does not imply causation.

Many techniques for carrying out regression analysis have been developed. Familiar methods such as linear regression and ordinary least squares regression are parametric, in that the regression function is defined in terms of a finite number of unknown parameters that are estimated from the data. Nonparametric regression refers to techniques that allow the regression function to lie in a specified set of functions, which may be infinite-dimensional.

The performance of regression analysis methods in practice depends on the form of the data generating process, and how it relates to the regression approach being used. Since the true form of the data-generating process is generally not known, regression analysis often depends to some extent on making assumptions about this process. These assumptions are sometimes testable if a sufficient quantity of data is available. Regression models for prediction are often useful even when the assumptions are moderately violated, although they may not perform optimally. However, in many applications, especially with small effects or questions of causality based on observational data, regression methods can give misleading results.

History

The earliest form of regression was the method of least squares, which was published by Legendre in 1805, and by Gauss in 1809. Legendre and Gauss both applied the method to the problem of determining, from astronomical observations, the orbits of bodies about the Sun (mostly comets, but also later the then newly discovered minor planets). Gauss published a further development of the theory of least squares in 1821, including a version of the Gauss–Markov theorem.

The term "regression" was coined by Francis Galton in the nineteenth century to describe a biological phenomenon. The phenomenon was that the heights of descendants of tall ancestors tend to regress down towards a normal average (a phenomenon also known as regression toward the mean). For Galton, regression had only this biological meaning, but his work was later extended by Udny Yule and Karl Pearson to a more general statistical context. In the work of Yule and Pearson, the joint distribution of the response and explanatory variables is assumed to be Gaussian. This assumption was weakened by R.A. Fisher in his works of 1922 and 1925. Fisher assumed that the conditional distribution of the response variable is Gaussian, but the joint distribution need not be. In this respect, Fisher's assumption is closer to Gauss's formulation of 1821.

In the 1950s and 1960s, economists used electromechanical desk calculators to calculate regressions. Before 1970, it sometimes took up to 24 hours to receive the result from one regression.

Regression methods continue to be an area of active research. In recent decades, new methods have been developed for robust regression, regression involving correlated responses such as time series and growth curves, regression in which the predictor or response variables are curves, images, graphs, or other complex data objects, regression methods accommodating various types of missing data, nonparametric regression, Bayesian methods for regression, regression in which the predictor variables are measured with error, regression with more predictor variables than observations, and causal inference with regression.

Regression Models

Regression models involve the following variables:

- The unknown parameters, denoted as β, which may represent a scalar or a vector.
- The independent variables, X.
- The dependent variable, Y.

In various fields of application, different terminologies are used in place of dependent and independent variables.

A regression model relates Y to a function of X and β.

$$Y \approx f(\mathbf{X}, \boldsymbol{\beta})$$

The approximation is usually formalized as $E(Y \mid X) = f(X, \beta)$. To carry out regression analysis, the form of the function f must be specified. Sometimes the form of this function is based on knowledge about the relationship between Y and X that does not rely on the data. If no such knowledge is available, a flexible or convenient form for f is chosen.

Assume now that the vector of unknown parameters β is of length k. In order to perform a regression analysis the user must provide information about the dependent variable Y:

- If N data points of the form (Y, X) are observed, where $N < k$, most classical approaches to regression analysis cannot be performed: since the system of equations defining the regression model is underdetermined, there are not enough data to recover β.
- If exactly $N = k$ data points are observed, and the function f is linear, the equations $Y = f(X, \beta)$ can be solved exactly rather than approximately. This reduces to solving a set of N equations with N unknowns (the elements of β), which has a unique solution as long as the X are linearly independent. If f is nonlinear, a solution may not exist, or many solutions may exist.
- The most common situation is where $N > k$ data points are observed. In this case, there is enough information in the data to estimate a unique value for β that best fits the data in some sense, and the regression model when applied to the data can be viewed as an overdetermined system in β.

In the last case, the regression analysis provides the tools for:

1. Finding a solution for unknown parameters β that will, for example, minimize the distance between the measured and predicted values of the dependent variable Y (also known as method of least squares).
2. Under certain statistical assumptions, the regression analysis uses the surplus of information to provide statistical information about the unknown parameters β and predicted values of the dependent variable Y.

Necessary Number of Independent Measurements

Consider a regression model which has three unknown parameters, β_0, β_1, and β_2. Suppose an experimenter performs 10 measurements all at exactly the same value of independent variable vector X (which contains the independent variables X_1, X_2, and X_3). In this case, regression analysis fails to give a unique set of estimated values for the three unknown parameters; the experimenter did not provide enough information. The best one can do is to estimate the average value and the standard deviation of the dependent variable Y. Similarly, measuring at two different values of X would give enough data for a regression with two unknowns, but not for three or more unknowns.

If the experimenter had performed measurements at three different values of the independent variable vector X, then regression analysis would provide a unique set of estimates for the three unknown parameters in β.

In the case of general linear regression, the above statement is equivalent to the requirement that the matrix X^TX is invertible.

Statistical Assumptions

When the number of measurements, *N*, is larger than the number of unknown parameters, *k*, and the measurement errors ε_i are normally distributed then *the excess of information* contained in (*N* " *k*) measurements is used to make statistical predictions about the unknown parameters. This excess of information is referred to as the degrees of freedom of the regression.

Underlying Assumptions

Classical assumptions for regression analysis include:

- The sample is representative of the population for the inference prediction.
- The error is a random variable with a mean of zero conditional on the explanatory variables.
- The independent variables are measured with no error. (Note: If this is not so, modelling may be done instead using errors-in-variables model techniques).
- The predictors are linearly independent, i.e. it is not possible to express any predictor as a linear combination of the others.
- The errors are uncorrelated, that is, the variance–covariance matrix of the errors is diagonal and each non-zero element is the variance of the error.
- The variance of the error is constant across observations (homoscedasticity). If not, weighted least squares or other methods might instead be used.

These are sufficient conditions for the least-squares estimator to possess desirable properties; in particular, these assumptions imply that the parameter estimates will be unbiased, consistent, and efficient in the class of linear unbiased estimators. It is important to note that actual data rarely satisfies the assumptions. That is, the method is used even though the assumptions are not true. Variation from the assumptions can sometimes be used as a measure of how far the model is from being useful. Many of these assumptions may be relaxed in more advanced treatments. Reports of statistical analyses usually include analyses of tests on the sample data and methodology for the fit and usefulness of the model.

Assumptions include the geometrical support of the variables. Independent and dependent variables often refer to values measured at point locations. There may be spatial trends and spatial autocorrelation in the variables that violate statistical assumptions of regression. Geographic weighted regression is one technique to deal with such data. Also, variables may include values aggregated by areas. With aggregated data the modifiable areal unit problem can cause extreme variation in regression parameters. When analyzing data aggregated by political boundaries, postal codes or census areas results may be very distinct with a different choice of units.

Linear Regression

In linear regression, the model specification is that the dependent variable, y_i is a linear combination of the *parameters* (but need not be linear in the *independent variables*). For example, in simple linear regression for modelling n data points there is one independent variable: x_i, and two parameters, β_0 and β_1:

$$\text{straight line: } y_i = \beta_0 + \beta_1 x_i + \varepsilon_i, \quad i = 1, \ldots, n.$$

In multiple linear regression, there are several independent variables or functions of independent variables.

Adding a term in x_i^2 to the preceding regression gives:

$$\text{parabola: } y_i = \beta_0 + \beta_1 x_i + \beta_2 x_i^2 + \varepsilon_i, i = 1, \ldots, n.$$

This is still linear regression; although the expression on the right hand side is quadratic in the independent variable x_i, it is linear in the parameters β_0, β_1 and

In both cases, ε_i is an error term and the subscript i indexes a particular observation.

Given a random sample from the population, we estimate the population parameters and obtain the sample linear regression model:

$$\widehat{y_i} = \hat{\beta}_0 + \hat{\beta}_1 x_i.$$

The residual, $e_i = y_i - \hat{y}_i$, is the difference between the value of the dependent variable predicted by the model, $\widehat{y_i}$, and the true value of the dependent variable, y_i. One method of estimation is ordinary least squares. This method obtains parameter estimates that minimize the sum of squared residuals, SSE, also sometimes denoted RSS:

$$SSE = \sum_{i=1}^{n} e_i^2.$$

Minimization of this function results in a set of normal equations, a set of simultaneous linear equations in the parameters, which are solved to yield the parameter estimators, .

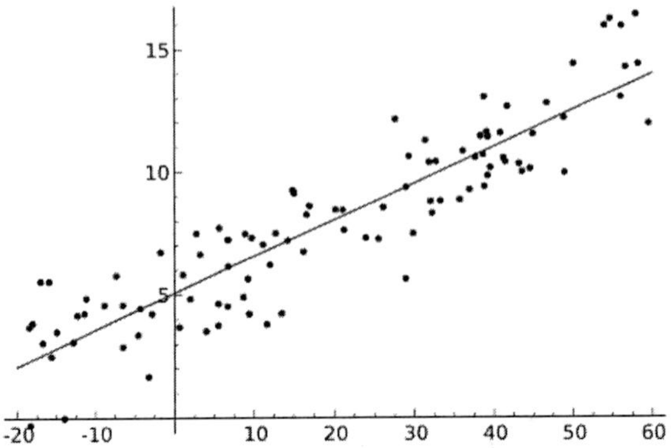

Figure: *Illustration of linear regression on a data set.*

In the case of simple regression, the formulas for the least squares estimates are

$$\widehat{\beta_1} = \frac{\sum (x_i - \overline{x})(y_i - \overline{y})}{\sum (x_i - \overline{x})^2} \text{ and } \widehat{\beta_0} = \overline{y} - \widehat{\beta_1}\overline{x}$$

where $\overline{x}$ is the mean (average) of the x values and $\overline{y}$ is the mean of the y values.

Under the assumption that the population error term has a constant variance, the estimate of that variance is given by:

$$\hat{\sigma}_\varepsilon^2 = \frac{SSE}{n-2}.$$

This is called the mean square error (MSE) of the regression. The denominator is the sample size reduced by the number of model parameters estimated from the same data, (n-p) for p regressors or (n-p-1) if an intercept is used. In this case, p=1 so the denominator is n-2.

The standard errors of the parameter estimates are given by

$$\hat{\sigma}_{\beta_0} = \hat{\sigma}_\varepsilon \sqrt{\frac{1}{n} + \frac{\overline{x}^2}{\sum (x_i - \overline{x})^2}}$$

$$\hat{\sigma}_{\beta_1} = \hat{\sigma}_\varepsilon \sqrt{\frac{1}{\sum (x_i - \overline{x})^2}}.$$

Under the further assumption that the population error term is normally distributed, the researcher can use these estimated standard errors to create confidence intervals and conduct hypothesis tests about the population parameters.

General Linear Model

In the more general multiple regression model, there are p independent variables:

$$y_i = \beta_1 x_{i1} + \beta_2 x_{i2} + \cdots + \beta_p x_{ip} + \varepsilon_i,$$

where x_{ij} is the ith observation on the jth independent variable, and where the first independent variable takes the value 1 for all i (so β_1 is the regression intercept).

The least squares parameter estimates are obtained from p normal equations. The residual can be written as

$$\varepsilon_i = y_i - \hat{\beta}_1 x_{i1} - \cdots - \hat{\beta}_p x_{ip}.$$

The normal equations are

$$\sum_{i=1}^{n}\sum_{k=1}^{p} X_{ij} X_{ik} \hat{\beta}_k = \sum_{i=1}^{n} X_{ij} y_i,\ j = 1, \ldots, p.$$

In matrix notation, the normal equations are written as

$$(\mathbf{X}^\top \mathbf{X})\hat{\boldsymbol{\beta}} = \mathbf{X}^\top \mathbf{Y},$$

where the ij element of X is x_{ij}, the i element of the column vector Y is y_i, and the j element of $\hat{\beta}$ is $\hat{\beta}_j$. Thus X is $n\times p$, Y is $n\times 1$, and $\hat{\beta}$ is $p\times 1$. The solution is

$$\hat{\boldsymbol{\beta}} = (\mathbf{X}^\top \mathbf{X})^{-1}\mathbf{X}^\top \mathbf{Y}.$$

Diagnostics

Once a regression model has been constructed, it may be important to confirm the goodness of fit of the model and the statistical significance of the estimated parameters. Commonly used checks of goodness of fit include the R-squared, analyses of the pattern of residuals and hypothesis testing. Statistical significance can be checked by an F-test of the overall fit, followed by t-tests of individual parameters.

Interpretations of these diagnostic tests rest heavily on the model assumptions. Although examination of the residuals can be used to invalidate a model, the results of a t-test or F-test are sometimes more difficult to interpret if the model's assumptions are violated. For example, if the error term does not have a normal distribution, in small samples the estimated parameters will not follow normal distributions and complicate inference. With relatively large samples, however, a central limit theorem can be invoked such that hypothesis testing may proceed using asymptotic approximations.

"Limited Dependent" Variables

The phrase "limited dependent" is used in econometric statistics for categorical and constrained variables.

The response variable may be non-continuous ("limited" to lie on some subset of the real line). For binary (zero or one) variables, if analysis proceeds with least-squares linear regression, the model is called the linear probability model. Nonlinear models for binary dependent variables include the probit and logit model. The multivariate probit model is a standard method of estimating a joint relationship between several binary dependent variables and some independent variables. For categorical variables with more than two values there is the multinomial logit. For ordinal variables with more than two values, there are the ordered logit and ordered probit models. Censored regression models may be used when the dependent variable is only sometimes observed, and Heckman correction type models may be used when the sample is not randomly selected from the population of interest. An alternative to such procedures is linear regression based on polychoric correlation (or polyserial correlations) between the categorical variables. Such procedures differ in the assumptions made about the distribution of the variables in the population. If the variable is positive with low values and represents the repetition of the occurrence of an event, then count models like the Poisson regression or the negative binomial model may be used instead.

Interpolation and Extrapolation

Regression models predict a value of the Y variable given known values of the X variables. Prediction *within* the range of values in the dataset used for model-fitting is known informally as interpolation. Prediction *outside* this range of the data is known as extrapolation. Performing extrapolation relies strongly on the regression assumptions. The further the extrapolation goes outside the data, the more room there is for the model to fail due to differences between the assumptions and the sample data or the true values. It is generally advised that when performing extrapolation, one should accompany the estimated value of the dependent variable with a prediction interval that represents the uncertainty. Such intervals tend to expand rapidly as the values of the independent variable(s) moved outside the range covered by the observed data. For such reasons and others, some tend to say that it might be unwise to undertake extrapolation. However, this does not cover the full set of modelling errors that may be being made: in particular, the assumption of a particular form for the relation between Y and X. A properly conducted regression analysis will include an assessment of how well the assumed form is matched by the observed data, but it can only do so within the range of values of the independent variables actually available. This means that any extrapolation is particularly reliant on the assumptions being made about the structural form of the regression relationship. Best-practice advice here is that a linear-in-variables and linear-in-parameters relationship should not be chosen simply for computational convenience, but that all available knowledge should be deployed in constructing a regression model. If this knowledge includes the fact that the dependent variable cannot go outside a certain range of values, this can be made use of in selecting the model – even if the observed dataset has no values particularly near such bounds. The implications of this step of choosing an appropriate functional form for the regression can be great when extrapolation is considered. At a minimum, it can ensure that any extrapolation arising from a fitted model is "realistic" (or in accord with what is known).

Nonlinear Regression

When the model function is not linear in the parameters, the sum of squares must be minimized by an iterative procedure. This introduces many complications which are summarized in Differences between linear and non-linear least squares

Power and Sample Size Calculations

There are no generally agreed methods for relating the number of observations versus the number of independent variables in the model. One rule of thumb suggested by Good and Hardin is $N = m^n$, where N is the sample size, n is the number of independent variables and m is the number of observations needed to reach the desired precision if the model had only one independent variable. For example, a researcher is building a linear regression model using a dataset that contains 1000 patients (N).

If the researcher decides that five observations are needed to precisely define a straight line (m), then the maximum number of independent variables the model can support is 4, because

$$\frac{\log 1000}{\log 5} = 4.29.$$

Other Methods

Although the parameters of a regression model are usually estimated using the method of least squares, other methods which have been used include:

- Bayesian methods, e.g. Bayesian linear regression
- Percentage regression, for situations where reducing *percentage* errors is deemed more appropriate.
- Least absolute deviations, which is more robust in the presence of outliers, leading to quantile regression
- Nonparametric regression, requires a large number of observations and is computationally intensive

Distance metric learning, which is learned by the search of a meaningful distance metric in a given input space.

Software

All major statistical software packages perform least squares regression analysis and inference. Simple linear regression and multiple regression using least squares can be done in some spreadsheet applications and on some calculators. While many statistical software packages can perform various types of nonparametric and robust regression, these methods are less standardized; different software packages implement different methods, and a method with a given name may be implemented differently in different packages. Specialised regression software has been developed for use in fields such as survey analysis and neuroimaging.

Chapter 8

Discriminant Function Analysis

Discriminant function analysis is a statistical analysis to predict a categorical dependent variable (called a grouping variable) by one or more continuous or binary independent variables (called predictor variables). The original dichotomous discriminant analysis was developed by Sir Ronald Fisher in 1936. It is different from an ANOVA or MANOVA, which is used to predict one (ANOVA) or multiple (MANOVA) continuous dependent variables by one or more independent categorical variables. Discriminant function analysis is useful in determining whether a set of variables is effective in predicting category membership.

Discriminant analysis is used when groups are known a priori (unlike in cluster analysis). Each case must have a score on one or more quantitative predictor measures, and a score on a group measure. In simple terms, discriminant function analysis is classification - the act of distributing things into groups, classes or categories of the same type.

Moreover, it is a useful follow-up procedure to a MANOVA instead of doing a series of one-way ANOVAs, for ascertaining how the groups differ on the composite of dependent variables. In this case, a significant F test allows classification based on a linear combination of predictor variables. Terminology can get confusing here, as in MANOVA, the dependent variables are the predictor variables, and the independent variables are the grouping variables.

Assumptions

The assumptions of discriminant analysis are the same as those for MANOVA. The analysis is quite sensitive to outliers and the size of the smallest group must be larger than the number of predictor variables.

- Multivariate normality: Independent variables are normal for each level of the grouping variable.
- Homogeneity of variance/covariance (homoscedasticity): Variances among group variables are the same across levels of predictors. Can be tested with Box's M statistic. It has been suggested, however, that linear discriminant analysis be used when covariances are equal, and that quadratic discriminant analysis may be used when covariances are not equal.
- Multicollinearity: Predictive power can decrease with an increased correlation between predictor variables.
- Independence: Participants are assumed to be randomly sampled, and a participant's score on one variable is assumed to be independent of scores on that variable for all other participants.

It has been suggested that discriminant analysis is relatively robust to slight violations of these assumptions, and it has also been shown that discriminant analysis may still be reliable when using dichotomous variables (where multivariate normality is often violated).

Discriminant Functions

Discriminant analysis works by creating one or more linear combinations of predictors, creating a new latent variable for each function. These functions are called discriminant functions. The number of functions possible is either *Ng*-1 where *Ng* = number of groups, or *p* (the number of predictors), whichever is smaller.

The first function created maximizes the differences between groups on that function. The second function maximizes differences on that function, but also must not be correlated with the previous function. This continues with subsequent functions with the requirement that the new function not be correlated with any of the previous functions.

Given group j, with $\mathbb{R}_j$ sets of sample space, there is a discriminant rule such that if $x \in \mathbb{R}_j$, then $x \in j$. Discriminant analysis then, finds "good" regions of $\mathbb{R}_j$ to minimize classification error, therefore leading to a high percent correct classified in the classification table. Hardle, W., Simar, L. (2007). *Applied Multivariate Statistical Analysis.* Springer Berlin Heidelberg.

Each function is given a discriminant score to determine how well it predicts group placement.

- Structure Correlation Coefficients: The correlation between each predictor and the discriminant score of each function. This is a whole correlation.Garson, G. D. (2008).
- Standardized Coefficients: Each predictor's unique contribution to each function, therefore this is a partial correlation. Indicates the relative importance of each predictor in predicting group assignment from each function.
- Functions at Group Centroids: Mean discriminant scores for each grouping variable are given for each function. The farther apart the means are, the less error there will be in classification.

Discrimination Rules

- Maximum likelihood: Assigns x to the group that maximizes population (group) density.
- Bayes Discriminant Rule: Assigns x to the group that maximizes $\pi_i f_i(x)$, where represents the prior probability of that classification, and π_i represents the population density.
- Fisher's linear discriminant rule: Maximizes the ratio between SS_{between} and SS_{within}, and finds a linear combination of the predictors to predict group.

Eigenvalues

An eigenvalue in discriminant analysis is the characteristic root of each function. It is an indication of how well that function differentiates the groups, where the larger the eigenvalue, the better the function differentiates. This however, should be interpreted with caution, as eigenvalues have no upper limit. The eigenvalue can be viewed as a ratio of SS_{between} and SS_{within} as in ANOVA when the dependent variable is the discriminant function, and the groups are the levels of the IV. This means that the largest eigenvalue is associated with the first function, the second largest with the second, etc..

Effect Size

Some suggest the use of eigenvalues as effect size measures, however, this is generally not supported. Instead, the canonical correlation is the preferred measure of effect size. It is similar to the eigenvalue, but is the square root of the ratio of SS_{between} and SS_{total}. It is the correlation between groups and the function. Another popular measure of effect size is the percent of variance for each function. This is calculated by: $(\lambda_x/\Sigma\lambda_i)$ X 100 where λ_x is the eigenvalue for the function and $\Sigma\lambda_i$ is the sum of all eigenvalues. This tells us how strong the prediction is for that particular function compared to the others. Percent correctly classified can also be analyzed as an effect size. The kappa value can describe this while correcting for chance agreement.

Variations

- Multiple discriminant analysis (MDA): related to MANOVA. Has more than two groups, and uses multiple dummy variables.
- Sequential discriminant analysis: assesses the importance of a set of IVs over and above a set of controls. In this case, the controls are entered first, and then the IVs.
- Stepwise discriminant analysis: Selects the most correlated predictor first, removes that variance in the grouping variable then adds the next most correlated and continues until the change in canonical correlation is not significant. Of course, both forward and backward stepwise procedures may be performed.

Comparison to Logistic Regression

Discriminant function analysis is very similar to logistic regression, and both can be used to answer the same research questions. Logistic regression does not have as many assumptions and restrictions as discriminant analysis.

However, when discriminant analysis' assumptions are met, it is more powerful than logistic regression.

Unlike logistic regression, discriminant analysis can be used with small sample sizes. It has been shown that when sample sizes are equal, and homogeneity of variance/covariance holds, discriminant analysis is more accurate. With all this being considered, logistic regression is the common choice nowadays, since the assumptions of discriminant analysis are rarely met.

Genome Evolution

Genome evolution is the process by which a genome changes in structure (sequence) or size over time. The study of genome evolution involves multiple fields such as structural analysis of the genome, the study of genomic parasites, gene and ancient genome duplications, polyploidy, and comparative genomics.

Genome evolution is a constantly changing and evolving field due to the steadily growing number of sequenced genomes, both prokaryotic and eukaryotic, available to the scientific community and the public at large.

History

Since the first sequenced genomes became available in the late 1970s, scientists have been using comparative genomics to study the differences and similarities between various genomes.

Genome sequencing has progressed over time to include more and more complex genomes including the eventual sequencing of the entire human genome in 2001.

By comparing genomes of both close relatives and distant ancestors the stark differences and similarities between species began to emerge as well as the mechanisms by which genomes are able evolve over time.

Prokaryotic and Eukaryotic Genomes

Prokaryotes

Prokaryotes are unicellular organisms that do not have membrane-bound organelles and lack a structurally distinct nucleus. Prokaryotic genomes have two main mechanisms of evolution: mutation and horizontal gene transfer.

A third mechanism, sexual reproduction, prominent in eukaryotes, is not found in bacteria. The large majority of prokaryotic genomes contain a single, circular chromosome. However, there are some bacteria that do possess more than one chromosome, such as *Vibrio cholera* which contains two circular chromosomes.

Other bacteria such as *Borrelia burgdorferi*, the causative bacterium in Lyme disease are unique in that they contain a linear chromosome as opposed to the more common circular conformation. The presence of only one chromosome makes prokaryotic organisms much more vulnerable to mutation.

Prokaryotes can acquire novel genetic material through the process of bacterial conjugation in which both plasmids and whole chromosomes can be passed between organisms. Plasmids are small extracellular and circular DNA. An often cited example of this process is the transfer of antibiotic resistance utilising plasmid DNA. Another mechanism of genome evolution is provided by transduction whereby bacteriophages introduce new DNA into a bacterial genome.

Genome evolution in bacteria is well understood because of the thousands of completely sequenced bacterial genomes available. Less is understood regarding eukaryotes since their genomes tend to be much larger and more complex making the sequencing of a eukaryotic genome much more costly and time-consuming in comparison.

**** E.Coli Largely Contains Only Exons in Genes. However, it Does Contain a Small Amount of Self-Splicing Introns (Group II).***

Characteristic	*E.Coli Genome*	*Human Genome*
Genome Size (base pairs)	4.6 Mb	3.2 Gb
Genome Structure	Circular	Linear
Number of chromosomes	1	46
Presence of Plasmids	Yes	No
Presence of Histones	No	Yes
DNA segregated in the nucleus	No	Yes
Number of genes	4,288	20,000
Presence of Introns	No*	Yes
Average Gene Size	700 bp	27,000 bp

Eukaryotes

Eukaryotic genomes are generally larger than that of the prokaryotes. While the *E.coli* genome is roughly 4.6Mb in length, in comparison the Human genome is much larger with a size of approximately 3.2Gb. The eukaryotic genome is linear and can be composed of multiple chromosomes, packaged in the nucleus of the cell. The non-coding portions of the gene, known as introns, which are largely not present in prokaryotes, are removed by RNA splicing before translation of the protein can occur. Eukaryotic genomes evolve over time through many mechanisms including sexual reproduction which introduces much greater genetic diversity to the offspring than the prokaryotic process of replication in which the offspring are theoretically genetic clones of the parental cell.

Genome Size

Genome size is usually measured in base pairs (or bases in single-stranded DNA or RNA). The C-value is another measure of genome size. Research on prokaryotic genomes shows that there is a significant positive correlation between the C-value of prokaryotes and the amount of genes that compose the genome. This indicates that gene number is the main factor influencing the size of the prokaryotic genome. In eukaryotic organisms, there is a paradox observed, namely that the number of genes that make up the genome does not correlate with genome size. In other words, the genome size is much larger than would be expected given the total number of protein coding genes.

Genome size can increase by duplication, insertion, or polyploidization. Recombination can lead to both DNA loss or gain.

Genomes can also shrink because of deletions. A famous example for such gene decay is the genome of *Mycobacterium leprae*, the causative agent of leprosy. *M.leprae* has lost many once-functional genes over time due to the formation of pseudogenes.

This is evident in looking at its closest ancestor *Mycobacterium tuberculosis*. *M. leprae* lives inside and replicates inside of a host and due to this arrangement it does not have a need for many of the genes it once carried which allowed it to live and prosper outside of the host. Thus over time these genes have lost their function through mechanisms such as mutation causing them to become pseudogenes. It is beneficial to an organism to rid itself of non-essential genes because it makes replicating its DNA much faster and requires less energy.

An example of increasing genome size over time is seen in filamentous plant pathogens. These plant pathogen genomes have been growing larger over the years due to repeat-driven expansion. The repeat-rich regions contain genes coding for host interaction proteins. With the addition of more and more repeats to these regions the plants increase the possibility of developing new virulence factors through mutation and other forms of genetic recombination. In this way it is beneficial for these plant pathogens to have larger genomes.

Mechanisms of Genome Evolution

Gene Duplication

Gene duplication is the process by which a region of DNA coding for a gene is duplicated. This can occur as the result of an error in recombination or through a retrotransposition event. Duplicate genes are often immune to the selective pressure under which genes normally exist. This can result in a large number of mutations accumulating in the duplicate gene code. This may render the gene non-functional or in some cases confer some benefit to the organism.

Whole Genome Duplication

Similar to gene duplication, whole genome duplication is the process by which an organism's entire genetic information is copied, once or multiple times which is known as polyploidy. This may provide an evolutionary benefit to the organism by supplying it with multiple copies of a gene thus creating a greater possibility of functional and selectively favoured genes.

In 1997, Wolfe & Shields gave evidence for an ancient duplication of the *Saccharomyces cerevisiae* (Yeast) genome. It was initially noted that this yeast genome contained many individual gene duplications.

Wolfe & Shields hypothesized that this was actually the result of an entire genome duplication in the yeast's distant evolutionary history. They found 32 pairs of homologous chromosomal regions, accounting for over half of the yeast's genome.

They also noted that although homologs were present, they were often located on different chromosomes. Based on these observations, they determined that *Saccharomyces cerevisiae* underwent a whole genome duplication soon after its evolutionary split from *Kluyveromyces*, a genus of ascomycetous yeasts. Over time, many of the duplicate genes were deleted and rendered non-functional.

A number of chromosomal rearrangements broke the original duplicate chromosomes into the current manifestation of homologous chromosomal regions. This idea was further solidified in looking at the genome of yeast's close relative *Ashbya gossypii.* Whole genome duplication is common in fungi as well as plant species. An example of extreme genome duplication is represented by the Common Cordgrass (*Spartina anglica*) which is a dodecaploid, meaning that it contains 12 sets of chromosomes, in stark contrast to the human diploid structure in which each individual has only two sets of 23 chromosomes.

Transposable Elements

Transposable elements are regions of DNA that can be inserted into the genetic code through one of two mechanisms. These mechanisms work similarly to "cut-and-paste" and "copy-and-paste" functionalities in word processing programs. The "cut-and-paste" mechanism works by excising DNA from one place in the genome and inserting itself into another location in the code. The "copy-and-paste" mechanism works by making a genetic copy or copies of a specific region of DNA and inserting these copies elsewhere in the code. The most common transposable element in the human genome is the Alu sequence, which is present in the genome over one million times.

Mutation

Spontaneous mutations often occur which can cause various changes in the genome. Mutations can result in the addition or deletion of one or more nucleotide bases. A change in the code can result in a frameshift mutation which causes the entire code to be read in the wrong order and thus often results in a protein becoming non-functional. A mutation in a promoter region, enhancer region or a region coding for transcription factors can also result in either a loss of function or and upregulation or downregulation in transcription of that gene.

Mutations are constantly occurring in an organism's genome and can cause either a negative effect, positive effect or no effect at all.

Pseudogenes

Often a result of spontaneous mutation, pseudogenes are dysfunctional genes derived from previously functional gene relatives. There are many mechanisms by which a functional gene can become a pseudogene including the deletion or insertion of one or multiple nucleotides. This can result in a shift of reading frame, causing the gene to longer code for the expected protein, a premature stop codon or a mutation in the promoter region. Oft cited examples of pseudogenes within the human genome include the once functional olfactory gene families. Over time, many olfactory genes in the human genome became pseudogenes and were no longer able to produce functional proteins, explaining the poor sense of smell humans possess in comparison to their mammalian relatives.

Exon Shuffling

Exon shuffling is a mechanism by which new genes are created. This can occur when two or more exons from different genes are combined together or when exons are duplicated. Exon shuffling results in new genes by altering the current intron-exon structure. This can occur by any of the following processes: transposon mediated shuffling, sexual recombination or illegitimate recombination. Exon shuffling may introduce new genes into the genome that can be either selected against and deleted or selectively favoured and conserved.

Genome Reduction and Gene Loss

Many species exhibit genome reduction when subsets of their genes are not needed anymore. This typically happens when organisms adapt to a parasitic life style, e.g. when their nutrients are supplied by a host. As a consequence, they lose the genes need to produce these nutrients. In many cases, there are both free living and parasitic species that can be compared and their lost genes identified. Good examples are the genomes of *Mycobacterium tuberculosis* and *Mycobacterium leprae*, the latter of which has a dramatically reduced genome.

Another beautiful example are endosymbiont species. For instance, *Polynucleobacter necessarius* was first described as a cytoplasmic endosymbiont of the ciliate *Euplotes aediculatus*. The latter species dies soon after being cured of the endosymbiont. In the few cases in which *P. necessarius* is not present, a different and rarer bacterium apparently supplies the same function. No attempt to grow symbiotic

P. necessarius outside their hosts has yet been successful, strongly suggesting that the relationship is obligate for both partners. Yet, closely related free-living relatives of P. necessarius have been identified. The endosymbionts have a significantly reduced genome when compared to their free-living relatives (1.56 Mbp vs. 2.16 Mbp).

Genome Evolution and Speciation

A major question of evolutionary biology is how genomes change to create new species. Speciation requires changes in behaviour, morphology, physiology, or metabolism (or combinations thereof). The evolution of genomes during speciation has been studied only very recently with the availability of next-generation sequencing technologies. For instance, cichlid fish in the three African lakes differ both morphologically and in their behaviour. The genomes of 5 species have revealed that both the sequences but also the expression pattern of many genes has quickly changed over a relatively short period of time (100,000 to several million years).

Notably, 20% of duplicate gene pairs have gained a completely new tissue-specific expression pattern, indicating that these genes also obtained new functions. Given that gene expression is driven by short regulatory sequences, this demonstrates that relatively few mutations are required to drive speciation. The cichlid genomes also showed increased evolutionary rates in microRNAs which are involved in gene expression.

Composition of Nucleotides (GC Content)

The genetic code is made up of sequences of four nucleotide bases: Adenine, Guanine, Cytosine and Thymine, commonly referred to as A,G,C,and T. The GC-content is the percentage of G & C bases within a genome. GC-content varies greatly between different organisms. Gene coding regions have been shown to have a higher GC-content and the longer the gene is, the greater the percentage of G and C bases that are present. A higher GC-content confers a benefit because a Guanine-Cytosine bond is made up of three hydrogen bonds while an Adenine-Thymine bond is made up of only two.

Thus the three hydrogen bonds give greater stability to the DNA strand. So, it is not surprising that important genes often have a higher GC-content than other parts of an organism's genome. For this reason, many species living at very high temperatures such as the ecosystems surrounding hydrothermal vents, have a very high GC-content. High GC-content is also seen in regulatory sequences such as promoters which signal the start of a gene. Many promoters contain CpG islands, areas of the genome where a cytosine nucleotide occurs next to a guanine nucleotide at a greater proportion. It has also been shown that a broad distribution of GC-content between species within a genus shows a more ancient ancestry. Since the species have had more time to evolve, their GC-content has diverged further apart.

Evolving Translation of Genetic Code

Amino acids are made up of three base long codons and both Glycine and Alanine are characterized by codons with Guanine-Cytosine bonds at the first two codon base positions. This GC bond gives more stability to the DNA structure.

It has been hypothesized that as the first organisms evolved in a high-heat and pressure environment they needed the stability of these GC bonds in their genetic code. There is a theory that our current understanding of the three base codon has evolved from what was once a two base code for amino acids.

De Novo Origin of Genes

Novel genes can arise from non-coding DNA. For instance, Levine and colleagues reported the origin of five new genes in the *D. melanogaster* genome from noncoding DNA. Subsequently, de novo origin of genes has been also shown in other organisms such as yeast, rice and humans.

Biochemical Systems Theory

Biochemical systems theory is a mathematical modelling framework for biochemical systems, based on ordinary differential equations (ODE), in which biochemical processes are represented using power-law expansions in the variables of the system.

This framework, which became known as Biochemical Systems Theory, has been developed since the 1960s by Michael Savageau and others for the systems analysis of biochemical processes. According to Cornish-Bowden (2007) they "regarded this as a general theory of metabolic control, which includes both metabolic control analysis and flux-oriented theory as special cases".

Representation

The dynamics of a species is represented by a differential equation with the structure:

$$\frac{dX_i}{dt} = \sum_j \mu_{ij} \cdot \gamma_j \prod_k X_k^{f_{jk}}$$

where X_i represents one of the n_d variables of the model (metabolite concentrations, protein concentrations or levels of gene expression). j represents the n_f biochemical processes affecting the dynamics of the species. On the other hand, μ_{ij} (stoichiometric coefficient), γ_j (rate constants) and f_{jk} (kinetic orders) are two different kinds of parameters defining the dynamics of the system.

The principal difference of power-law models with respect to other ODE models used in biochemical systems is that the kinetic orders can be non-integer numbers. A kinetic order can have even negative value when inhibition is modelled. In this way, power-law models have a higher flexibility to reproduce the non-linearity of biochemical systems.

Models using power-law expansions have been used during the last 35 years to model and analyse several kinds of biochemical systems including metabolic networks, genetic networks and recently in cell signalling.

Functional Genomics

Functional genomics is a field of molecular biology that attempts to make use of the vast wealth of data produced by genomic projects (such as genome sequencing projects) to describe gene (and protein) functions and interactions. Unlike genomics, functional genomics focuses on the dynamic aspects such as gene transcription, translation, and protein–protein interactions, as opposed to the static aspects of the genomic information such as DNA sequence or structures. Functional genomics attempts to answer questions about the function of DNA at the levels of genes, RNA transcripts, and protein products. A key characteristic of functional genomics studies is their genome-wide approach to these questions, generally involving high-throughput methods rather than a more traditional "gene-by-gene" approach.

Goals of Functional Genomics

The goal of functional genomics is to understand the relationship between an organism's genome and its phenotype. The term functional genomics is often used broadly to refer to the many possible approaches to understanding the properties and function of the entirety of an organism's genes and gene products. This definition is somewhat variable; Gibson and Muse define it as "approaches under development to ascertain the biochemical, cellular, and/or physiological properties of each and every gene product", while Pevsner includes the study of nongenic elements in his definition: "the genome-wide study of the function of DNA (including genes and nongenic elements), as well as the nucleic acid and protein products encoded by DNA". Functional genomics involves studies of natural variation in genes, RNA, and proteins over time (such as an organism's development) or space (such as its body regions), as well as studies of natural or experimental functional disruptions affecting genes, chromosomes, RNA, or proteins.

The promise of functional genomics is to expand and synthesize genomic and proteomic knowledge into an understanding of the dynamic properties of an organism at cellular and/or organismal levels. This would provide a more complete picture of how biological function arises from the information encoded in an organism's genome. The possibility of understanding how a particular mutation leads to a given phenotype has important implications for human genetic diseases, as answering these questions could point scientists in the direction of a treatment or cure.

Techniques and Applications

Functional genomics includes function-related aspects of the genome itself such as mutation and polymorphism (such as single nucleotide polymorphism (SNP) analysis), as well as measurement of molecular activities. The latter comprise a number of "-omics" such as transcriptomics (gene expression), proteomics (protein expression), and metabolomics. Functional genomics uses mostly multiplex techniques to measure the abundance of many or all gene products such as mRNAs or proteins within a biological sample. Together these measurement modalities endeavour to quantitate the various biological processes and improve our understanding of gene and protein functions and interactions.

At the DNA Level

Genetic Interaction Mapping

Systematic pairwise deletion of genes or inhibition of gene expression can be used to identify genes with related function, even if they do not interact physically. Epistasis refers to the fact that effects for two different gene knockouts may not be additive; that is, the phenotype that results when two genes are inhibited may be different from the sum of the effects of single knockouts.

The ENCODE Project

The ENCODE (Encyclopedia of DNA elements) project is an in-depth analysis of the human genome whose goal is to identify all the functional elements of genomic DNA, in both coding and noncoding regions. To this point, only the pilot phase of the study has been completed, involving hundreds of assays performed on 44 regions of known or unknown function comprising 1% of the human genome. Important results include evidence from genomic tiling arrays that most nucleotides are transcribed as coding transcripts, noncoding RNAs, or random transcripts, the discovery of additional transcriptional regulatory sites, further elucidation of chromatin-modifying mechanisms.

At the RNA Level: Transcriptome Profiling

Microarrays

Microarrays measure the amount of mRNA in a sample that corresponds to a given gene or probe DNA sequence. Probe sequences are immobilized on a solid surface and allowed to hybridize with fluorescently labelled "target" mRNA. The intensity of fluorescence of a spot is proportional to the amount of target sequence that has hybridized to that spot, and therefore to the abundance of that mRNA sequence in the sample. Microarrays allow for identification of candidate genes involved in a given process based on variation between transcript levels for different conditions and shared expression patterns with genes of known function.

SAGE

SAGE (serial analysis of gene expression) is an alternate method of gene expression analysis based on RNA sequencing rather than hybridization. SAGE relies on the sequencing of 10–17 base pair tags which are unique to each gene. These tags are produced from poly-A mRNA and ligated end-to-end before sequencing. SAGE gives an unbiased measurement of the number of transcripts per cell, since it does not depend on prior knowledge of what transcripts to study (as microarrays do).

At the Protein Level: Protein–Protein Interactions

Yeast Two-Hybrid System

A yeast two-hybrid (Y2H) screen tests a "bait" protein against many potential interacting proteins ("prey") to identify physical protein–protein interactions. This system is based on a transcription factor, originally GAL4, whose separate DNA-binding and transcription activation domains are both required in order for the protein to cause transcription of a reporter gene. In a Y2H screen, the "bait" protein is fused to the binding domain of GAL4, and a library of potential "prey" (interacting) proteins is recombinantly expressed in a vector with the activation domain. In vivo interaction of bait and prey proteins in a yeast cell brings the activation and binding domains of GAL4 close enough together to result in expression of a reporter gene. It is also possible to systematically test a library of bait proteins against a library of prey proteins to identify all possible interactions in a cell.

AP/MS

Affinity purification and mass spectrometry (AP/MS) is able to identify proteins that interact with one another in complexes. Complexes of proteins are allowed to form around a particular "bait" protein. The bait protein is identified using an antibody or a recombinant tag which allows it to be extracted along with any proteins that have formed a complex with it. The proteins are then digested into short peptide fragments and mass spectrometry is used to identify the proteins based on the mass-to-charge ratios of those fragments.

Loss-of-Function Techniques

Mutagenesis

Gene function can be investigated by systematically "knocking out" genes one by one. This is done by either deletion or disruption of function (such as by insertional mutagenesis) and the resulting organisms are screened for phenotypes that provide clues to the function of the disrupted gene.

RNAi

RNA interference (RNAi) methods can be used to transiently silence or knock down gene expression using ~20 base-pair double-stranded RNA typically delivered by transfection of synthetic ~20-mer short-interfering RNA molecules (siRNAs) or by virally encoded short-hairpin RNAs (shRNAs). RNAi screens, typically performed in cell culture-based assays or experimental organisms (such as C. elegans) can be used to systematically disrupt nearly every gene in a genome or subsets of genes (sub-genomes); possible functions of disrupted genes can be assigned based on observed phenotypes.

Functional Annotations for Genes

Genome Annotation

Putative genes can be identified by scanning a genome for regions likely to encode proteins, based on characteristics such as long open reading frames, transcriptional initiation sequences, and polyadenylation sites. A sequence identified as a putative gene must be confirmed by further evidence, such as similarity to cDNA or EST sequences from the same organism, similarity of the predicted protein sequence to known proteins, association with promoter sequences, or evidence that mutating the sequence produces an observable phenotype.

Rosetta Stone Approach

The Rosetta stone approach is a computation method of de novo protein function prediction, based on the hypothesis that some proteins involved in a given physiological process may exist as two separate genes in one organism and as a single gene in another. Genomes are scanned for sequences that are independent in one organism and in a single open reading frame in another. If two genes have fused, it is predicted that they have similar biological functions that make such coregulation advantageous.

Functional Genomics and Bioinformatics

Because of the large quantity of data produced by these techniques and the desire to find biologically meaningful patterns, bioinformatics is crucial to analysis of functional genomics data. Examples of techniques in this class are data clustering or principal component analysis for unsupervised machine learning (class detection) as well as artificial neural networks or support vector machines for supervised machine learning (class prediction, classification). Functional enrichment analysis is used to determine the extent of over- or under-expression (positive- or negative- regulators in case of RNAi screens) of functional categories relative to a background sets. Gene ontology based enrichment analysis are provided by DAVID and Gene Set Enrichment Analysis (GSEA), pathway based analysis by Ingenuity and Pathway studio and protein complex based analysis by COMPLEAT.

Structural Genomics

Structural genomics seeks to describe the 3-dimensional structure of every protein encoded by a given genome. This genome-based approach allows for a high-throughput method of structure determination by a combination of experimental and modelling approaches. The principal difference between structural genomics and traditional structural prediction is that structural genomics attempts to determine the structure of every protein encoded by the genome, rather than focusing on one particular protein. With full-genome sequences available, structure prediction can be done more quickly through a combination of experimental and modelling approaches, especially because the availability of large number of sequenced genomes and previously solved protein structures allows scientists to model protein structure on the structures of previously solved homologs.

Because protein structure is closely linked with protein function, the structural genomics has the potential to inform knowledge of protein function. In addition to elucidating protein functions, structural genomics can be used to identify novel protein folds and potential targets for drug discovery. Structural genomics involves taking a large number of approaches to structure determination, including experimental methods using genomic sequences or modelling-based approaches based on sequence or structural homology to a protein of known structure or based on chemical and physical principles for a protein with no homology to any known structure.

As opposed to traditional structural biology, the determination of a protein structure through a structural genomics effort often (but not always) comes before anything is known regarding the protein function. This raises new challenges in structural bioinformatics, i.e. determining protein function from its 3D structure.

Structural genomics emphasizes high throughput determination of protein structures. This is performed in dedicated centres of structural genomics.

While most structural biologists pursue structures of individual proteins or protein groups, specialists in structural genomics pursue structures of proteins on a genome wide scale. This implies large-scale cloning, expression and purification. One main advantage of this approach is economy of scale. On the other hand, the scientific value of some resultant structures is at times questioned. A *Science* article from January 2006 analyzes the structural genomics field.

One advantage of structural genomics, such as the Protein Structure Initiative, is that the scientific community gets immediate access to new structures, as well as to reagents such as clones and protein. A disadvantage is that many of these structures are of proteins of unknown function and do not have corresponding publications. This requires new ways of communicating this structural information to the broader research community.

The Bioinformatics core of the Joint centre for structural genomics (JCSG) has recently developed a wiki-based approach namely Open protein structure annotation network (TOPSAN) (link) for annotating protein structures emerging from high-throughput structural genomics centres.

Goals

One goal of structural genomics is to identify novel protein folds. Experimental methods of protein structure determination require proteins that express and/or crystallize well, which may inherently bias the kinds of proteins folds that this experimental data elucidate. A genomic, modelling-based approach such as *ab initio* modelling may be better able to identify novel protein folds than the experimental approaches because they are not limited by experimental constraints.

Protein function depends on 3-D structure and these 3-D structures are more highly conserved than sequences. Thus, the high-throughput structure determination methods of structural genomics have the potential to inform our understanding of protein functions. This also has potential implications for drug discovery and protein engineering. Furthermore, every protein that is added to the structural database increases the likelihood that the database will include homologous sequences of other unknown proteins.

The Protein Structure Initiative (PSI) is a multifaceted effort funded by the National Institutes of Health with various academic and industrial partners that aims to increase knowledge of protein structure using a structural genomics approach and to improve structure-determination methodology.

Methods

Structural genomics takes advantage of completed genome sequences in several ways in order to determine protein structures. The gene sequence of the target protein can also be compared to a known sequence and structural information can then be inferred from the known protein's structure. Structural genomics can be used to predict novel protein folds based on other structural data. Structural genomics can also take modelling-based approach that relies on homology between the unknown protein and a solved protein structure.

de Novo *Methods*

Completed genome sequences allow every open reading frame (ORF), the part of a gene that is likely to contain the sequence for the mRNA and protein, to be cloned and expressed as protein. These proteins are then purified and crystallized, and then subjected to one of two types of structure determination: X-ray crystallography and Nuclear Magnetic Resonance (NMR).

The whole genome sequence allows for the design of every primer required in order to amplify all of the ORFs, clone them into bacteria, and then express them. By using a whole-genome approach to this traditional method of protein structure determination, all of the proteins encoded by the genome can be expressed at once. This approach allows for the structural determination of every protein that is encoded by the genome.

Modelling-Based Methods

Ab Initio *Modelling*

This approach uses protein sequence data and the chemical and physical interactions of the encoded amino acids to predict the 3-D structures of proteins with no homology to solved protein structures. One highly successful method for *ab initio* modelling is the Rosetta program, which divides the protein into short segments and arranges short polypeptide chain into a low-energy local conformation. Rosetta is available for commercial use and for non-commercial use through its public program, Robetta.

Sequence-based Modelling

This modelling technique compares the gene sequence of an unknown protein with sequences of proteins with known structures. Depending on the degree of similarity between the sequences, the structure of the known protein can be used as a model for solving the structure of the unknown protein. Highly accurate modelling is considered to require at least 50% amino acid sequence identity between the unknown protein and the solved structure. 30-50% sequence identity gives a model of intermediate-accuracy, and sequence identity below 30% gives low-accuracy models. It has been predicted that at least 16,000 protein structures will need to be determined in order for all structural motifs to be represented at least once and thus allowing the structure of any unknown protein to be solved accurately through modelling. One disadvantage of this method, however, is that structure is more conserved than sequence and thus sequence-based modelling may not be the most accurate way to predict protein structures.

Threading

Threading bases structural modelling on fold similarities rather than sequence identity. This method may help identify distantly related proteins and can be used to infer molecular functions.

Examples of Structural Genomics

There are currently a number of on-going efforts to solve the structures for every protein in a given proteome.

The Thermotogo Maritima *Proteome*

One current goal of the Joint Centre for Structural Genomics (JCSG), a part of the Protein Structure Initiative (PSI) is to solve the structures for all the proteins in *Thermotogo maritima*, a thermophillic bacterium. *T. maritima* was selected as a structural genomics target based on its relatively small genome consisting of 1,877 genes and the hypothesis that the proteins expressed by a thermophilic bacterium would be easier to crystallize.

Lesley *et al* used *Escherichia coli* to express all the open-reading frames (ORFs) of *T. martima*. These proteins were then crystallized and structures were determined for successfully crystallized proteins using X-ray crystallography. Among other structures, this structural genomics approach allowed for the determination of the structure of the TM0449 protein, which was found to exhibit a novel fold as it did not share structural homology with any known protein.

The Mycobacterium Tuberculosis *Proteome*

The goal of the TB Structural Genomics Consortium is to determine the structures of potential drug targets in *Mycobacterium tuberculosis*, the bacterium that causes tuberculosis. The development of novel drug therapies against tuberculosis are particularly important given the growing problem of multi-drug-resistant tuberculosis.

The fully sequenced genome of *M. tuberculosis* has allowed scientists to clone many of these protein targets into expression vectors for purification and structure determination by X-ray crystallography. Studies have identified a number of target proteins for structure determination, including extracellular proteins that may be involved in pathogenesis, iron-regulatory proteins, current drug targets, and proteins predicted to have novel folds. So far, structures have been determined for 708 of the proteins encoded by *M. tuberculosis*.

Gene Ontology

Gene ontology, or GO, is a major bioinformatics initiative to unify the representation of gene and gene product attributes across all species. More specifically, the project aims to:

1. Maintain and develop its controlled vocabulary of gene and gene product attributes;
2. Annotate genes and gene products, and assimilate and disseminate annotation data;
3. Provide tools for easy access to all aspects of the data provided by the project, and to enable functional interpretation of

experimental data using the GO, for example via enrichment analysis.

GO is part of a larger classification effort, the Open Biomedical Ontologies (OBO).

GO Terms and Ontology

From a practical view, an ontology is a representation of something we know about. "Ontologies" consist of a representation of things that are detectable or directly observable, and the relationships between those things. There is no universal standard terminology in biology and related domains, and term usages may be specific to a species, research area or even a particular research group. This makes communication and sharing of data more difficult. The Gene Ontology project provides an ontology of defined terms representing gene product properties. The ontology covers three domains:

- cellular component, the parts of a cell or its extracellular environment;
- molecular function, the elemental activities of a gene product at the molecular level, such as binding or catalysis;
- biological process, operations or sets of molecular events with a defined beginning and end, pertinent to the functioning of integrated living units: cells, tissues, organs, and organisms.

Each GO term within the ontology has a term name, which may be a word or string of words; a unique alphanumeric identifier; a definition with cited sources; and a namespace indicating the domain to which it belongs. Terms may also have synonyms, which are classed as being exactly equivalent to the term name, broader, narrower, or related; references to equivalent concepts in other databases; and comments on term meaning or usage. The GO ontology is structured as a directed acyclic graph, and each term has defined relationships to one or more other terms in the same domain, and sometimes to other domains. The GO vocabulary is designed to be species-neutral, and includes terms applicable to prokaryotes and eukaryotes, single and multicellular organisms.

GO is not static, and additions, corrections and alterations are suggested by, and solicited from, members of the research and annotation communities, as well as by those directly involved in the GO project. For example, an annotator may request a specific term to represent a metabolic pathway, or a section of the ontology may be revised with the help of community experts (e.g.). Suggested edits are reviewed by the ontology editors, and implemented where appropriate.

The GO ontology file is freely available from the GO website in a number of formats, or can be accessed online using the GO browser AmiGO.

Example GO Term

```
id:         GO:0000016
name:       lactase activity
namespace:  molecular_function
def:        "Catalysis of the reaction: lactose + H2O = D-glucose + D-
galactose." [EC:3.2.1.108]
synonym:    "lactase-phlorizin hydrolase activity" BROAD [EC:3.2.1.108]
synonym:    "lactose galactohydrolase activity" EXACT [EC:3.2.1.108]
xref:       EC:3.2.1.108
xref:       MetaCyc:LACTASE-RXN
xref:       Reactome:20536
is_a:        GO:0004553 ! hydrolase activity, hydrolyzing O-glycosyl
compounds
```

Annotation

Genome annotation is the practice of capturing data about a gene product, and GO annotations use terms from the GO ontology to do so. The members of the GO Consortium submit their annotation for integration and dissemination on the GO website, where they can be downloaded directly or viewed online using AmiGO. In addition to the gene product identifier and the relevant GO term, GO annotations have the following data:

- The *reference* used to make the annotation (e.g. a journal article)
- An *evidence code* denoting the type of evidence upon which the annotation is based
- The date and the creator of the annotation

The evidence code comes from the Evidence Code Ontology, a controlled vocabulary of codes covering both manual and automated annotation methods. For example, *Traceable Author Statement* (TAS) means a curator has read a published scientific paper and the metadata for that annotation bears a citation to that paper; *Inferred from Sequence Similarity* (ISS) means a human curator has reviewed the output from a sequence similarity search and verified that it is biologically meaningful.

Annotations from automated processes (for example, remapping annotations created using another annotation vocabulary) are given the code *Inferred from Electronic Annotation* (IEA). As of April 1, 2010, over 98% of all GO annotations were inferred computationally, not by curators. As these annotations are not checked by a human, the GO Consortium considers them to be less reliable and includes only a subset in the data available online in AmiGO. Full annotation data sets can be downloaded from the GO website. To support the development of annotation, the GO Consortium provides study camps and mentors to new groups of developers.

Example Annotation

```
Gene product:     Actin, alpha cardiac muscle 1, UniProtKB:P68032
GO term:          heart contraction ; GO:0060047 (biological process)
Evidence code:    Inferred from Mutant Phenotype (IMP)
Reference:        PMID 17611253
Assigned by:        UniProtKB, June 6, 2008
```

Tools

There are a large number of tools available both online and to download that use the data provided by the GO project. The vast majority of these come from third parties; the GO Consortium develops and supports two tools, AmiGO and OBO-Edit.

AmiGO is a web-based application that allows users to query, browse and visualize ontologies and gene product annotation data. In addition, it also has a BLAST tool, tools allowing analysis of larger data sets, and an interface to query the GO database directly.

AmiGO can be used online at the GO website to access the data provided by the GO Consortium, or can be downloaded and installed for local use on any database employing the GO database schema (e.g.). It is free open source software and is available as part of the go-dev software distribution.

OBO-Edit is an open source, platform-independent ontology editor developed and maintained by the Gene Ontology Consortium. It is implemented in Java, and uses a graph-oriented approach to display and edit ontologies. OBO-Edit includes a comprehensive search and filter interface, with the option to render subsets of terms to make them visually distinct; the user interface can also be customized according to user preferences. OBO-Edit also has a reasoner that can infer links that have not been explicitly stated, based on existing relationships and their properties. Although it was developed for biomedical ontologies, OBO-Edit can be used to view, search and edit any ontology. It is freely available to download.

GO Consortium

The GO Consortium is the set of biological databases and research groups actively involved in the GO project. This includes a number of model organism databases and multi-species protein databases, software development groups, and a dedicated editorial office.

History

Gene ontology was originally constructed in 1998 by a consortium of researchers studying the genome of three model organisms: *Drosophila melanogaster* (fruit fly), *Mus musculus* (mouse), and *Saccharomyces cerevisiae* (brewer's or baker's yeast). Many other model organism databases have joined the Gene Ontology consortium, contributing not only annotation data, but also contributing to the development of the ontologies and tools to view and apply the data. Until now, most of major databases in plant, animal and microorganism make a contribution towards this project.

As of January 2008, GO contains over 24,500 terms applicable to a wide variety of biological organisms. There is a significant body of literature on the development and use of GO, and it has become a standard tool in the bioinformatics arsenal. Their objectives have three aspects: building gene ontology, assigning ontology to gene/gene products and developing software and databases for the first two objects.

Bibliography

Aluru, Srinivas: *Handbook of Computational Molecular Biology, Chapman & Hall/Crc*, United Kingdom, 2006.

Baldi, P. and Brunak, S.: *Bioinformatics: The Machine Learning Approach*, MIT Press, United States, 2001.

Barnes, M.R. and Gray, I.C.: *Bioinformatics for Geneticists*, Wiley, New York, 2003.

Baxevanis, A.D. and Ouellette, B.F.F.: *Bioinformatics: A Practical Guide to the Analysis of Genes and Proteins*, Wiley, New York, 2005.

Baxevanis, A.D., Petsko, G.A., Stein, L.D., and Stormo, G.D.: *Current Protocols in Bioinformatics*, Wiley, New York, 2007.

Bergman NH.: *Comparative Genomics*, Humana Press, Totowa (NJ), 2007.

Burks, Arthur W.: *Chance, Cause and Reason: An Inquiry into the Nature of Scientific Evidence*, University of Chicago Press, 1978.

Cox, Richard Threlkeld: *The algebra of probable inference*, Johns Hopkins Press, Baltimore, 1961.

Cristianini, N. and Hahn, M.: *Introduction to Computational Genomics*, Cambridge University Press, 2006.

Durbin, R., S. Eddy, A. Krogh and G. Mitchison: *Biological sequence analysis*. Cambridge University Press, 1998.

Finetti, Bruno: *Foresight: its Logical laws, its Subjective Sources*, Wiley, New York, 1964.

Fristedt, Bert; Gray, Lawrence: *A modern approach to probability theory*, Birkhäuser, Boston, 1996.

George, Casella and Roger, L. Berger: *Statistical Inference*, Duxbury Press, USA, 1990.

Hazewinkel, Michiel: *Random variable*, Springer, New York, 2001.

Jaynes, E. T.: *Probability theory the logic of science*, Cambridge University Press, Cambridge, UK New York, NY, 2003.

Kallenberg, Olav: *Random Measures*, Akademie Verlag, Berlin, 1986.

Keedwell, E.: *Intelligent Bioinformatics: The Application of Artificial Intelligence Techniques to Bioinformatics Problems,* Wiley, New York, 2005.

Kohane: *Microarrays for an Integrative Genomics,* The MIT Press, Cambridge, Massachusetts, United States, 2002.

Laplace, P. S.: *A Philosophical Essay on Probabilities,* Dover Publications, New York, 1814.

Lund, O.: *Immunological Bioinformatics,* The MIT Press, Cambridge, Massachusetts, United States, 2005.

Mises, Richard: *Probability, statistics, and truth,* Dover Publications, New York, 1981.

Pachter, Lior and Sturmfels, Bernd: *Algebraic Statistics for Computational Biology,* Cambridge University Press, New York, 1981.

Papoulis, Athanasios: *Probability, Random Variables, and Stochastic Processes,* McGraw–Hill, Tokyo, 1965.

Paulos, J.A.: *Innumeracy: Mathematical Illiteracy and its Consequences,* Hill and Wang, United States, 1988.

Peterson, Martin: *An introduction to decision theory,* Cambridge University Press, Cambridge, UK New York, 2009.

Pevzner, Pavel A.: *Computational Molecular Biology: An Algorithmic Approach,* The MIT Press, Cambridge, United States, 2000.

Richard Durbin, Sean R. Eddy, Anders Krogh, Graeme Mitchison: *Biological Sequence Analysis: Probabilistic Models of Proteins and Nucleic Acids,* Cambridge University Press, 1999.

Savage, L.J.: *The foundations of statistics,* John Wiley & Sons, New York, 1954.

Sheldon Ross: *A First Course in Probability,* Pearson Prentice Hall, United States, 2010

Spanos, Aris: *Statistical foundations of econometric modelling,* Cambridge University Press, Cambridge New York, 1986.

Stevens, Hallam: *Life Out of Sequence: A Data-Driven History of Bioinformatics,* The University of Chicago Press, Chicago, Cambridge New York, 1986.

Tisdall, James: *Beginning Perl for Bioinformatics,* O'Reilly, United States, 2001.

Xuedong Huang, Alex Acero, and Hsiao-Wuen Hon: *Spoken Language Processing,* Prentice Hall, New York, 2001.

Xuedong Huang, M. Jack, and Y. Ariki: *Hidden Markov Models for Speech Recognition,* Edinburgh University Press, 1990.

Yates, Daniel S.; Moore, David S; Starnes, Daren S.: *The Practice of Statistics,* Freeman, New York, 2003.

Index

L

M

N

O

P

R

S

T

V

W